Radio and Microwave Spectroscopy

DAVID J.E. INGRAM

M.A., D.Sc., Hon. D.Sc.
Principal, Chelsea College,
University of London

A HALSTED PRESS BOOK

JOHN WILEY & SONS

New York—Toronto

English edition first published in 1976 by
Butterworth & Co. (Publishers) Ltd
88 Kingsway, London WC2B 6AB

Published in the U.S.A. and Canada by
Halsted Press, a Division of John Wiley & Sons Inc.,
New York

First published 1976 — **82738**

ISBN 470–42788–4
LC   75–8317

Printed and bound in Great Britain by R. J. Acford Ltd, Chichester, Sussex.

Preface

Like other branches of spectroscopy before it, spectroscopy in the microwave and radio regions of the spectrum has now moved on from the research area into normal undergraduate teaching courses in both physics and chemistry. Moreover, very considerable general interest now exists in the different forms of spectroscopy in these two new regions, since they not only have a large number of important applications, but also serve to illustrate in a very striking way some of the basic principles in physics and chemistry. For both of these reasons it was felt that the time was opportune for an undergraduate text on this topic to put this whole area in the perspective appropriate to the rest of a physics or chemistry course.

The book has therefore been written for both physicists and chemists, at about their second year of study, and has been designed to fit in with either single or joint honours courses in the physical sciences area. Its length has been limited, moreover, so that it can also be used as a text for the unit course structures which are now being adopted for physics and chemistry in the United Kingdom, United States of America and elsewhere. In view of this, any detailed high-level mathematical treatment has been avoided and stress has been placed instead on the underlying physical principles. For this reason, the book should also serve as a readable introduction for postgraduate students and staff in other subjects, such as biology and electronics, since these regions of spectroscopy are now extending into a wide variety of other sciences. The integration between the sciences, both in the way of design of instrumentation as well as in its applications, is one of the main themes of this book, and for this reason it is hoped it will also be of general interest in the last years at school.

Although reference is made to standard results obtained by wave-mechanics or simple quantum mechanics, the mathematical ability of the reader himself is not assumed to be beyond that of 'A' level, or the specialist school leaver, so that the text can be understood and followed through by any physics or chemistry undergraduate. Moreover, an attempt has been made to introduce each new topic in as qualitative way as possible first, so that the basic principles can be clearly understood before any detailed treatment is undertaken. Thus, the first chapter considers the place of the radio and microwave regions in the complete range of the electromagnetic spectrum, and then concentrates on particular features and characteristics which belong to this wavelength region. Following chapters then consider the different types of spectroscopy that have developed, the one on the straightforward spectroscopy of gases being treated first. The extension of this into solid state studies and the various applications of electron resonance at microwave frequencies are then described in the third chapter, this being followed in turn by a chapter devoted to nuclear magnetic resonance at radio-frequencies.

One of the fascinating features of spectroscopy at these frequencies is the interplay that has taken place continually between pure and applied science. Thus, all the techniques which were necessary to open up these new regions of the spectrum came from the concentrated efforts of the applied electronic engineers when developing radar in the last war. These were then taken up by the pure scientists in university research laboratories to initiate the new forms of spectroscopy considered in this book. From these studies have come, in turn, other applications of intense interest and importance to the applied scientist and engineer, such as new standards of frequency and time, and devices such as the maser and laser. These various applications to both pure and applied science are summarised in the last chapters, and serve as a very good example of how the 'wheel has turned full circle', and the pure scientists have repaid to some extent the debt they owed to the applied engineers. In this way, I believe the topics treated in this book serve as a very good example of the way in which science should progress with a constant interplay between the pure and applied scientist on the one hand, together with an increasing interaction between the different scientific disciplines, on the other.

Maybe I can close on a slight personal note and say that it is interesting to me to see that exactly twenty years have elapsed between the publication of this text and one written to introduce research workers to the new subject of 'Spectroscopy at Radio and Microwave Frequencies'. Maybe this is just about the right phase lag between the initiation of an exciting new research field and the assimilation of that work into a typical undergraduate course!

David J.E. Ingram

Chelsea College,
University of London

Contents

CONTENTS

Chapter One

The Radio and Microwave Regions of the Spectrum

1.1 The Complete Electromagnetic Spectrum

Spectroscopy at both radio and microwave frequencies has now become established as a standard tool which is employed to study the structure of atoms and molecules in the gaseous, liquid and solid states and has thus now taken its place with the earlier types of spectroscopy which cover the other regions of the electromagnetic spectrum.

It is interesting in this connection to see how knowledge of the different regions of the electromagnetic spectrum, and the different radiations within it, has always developed in step with a growing understanding of the structure of matter itself, and of the forces and interactions between its individual components. In this way the history of new ideas and techniques in spectroscopy mirrors the history of the development of physics itself. This integration of knowledge can be illustrated in a schematic way in *Figure 1.1* where the different types of radiation which form the complete electromagnetic spectrum are drawn out horizontally across the centre of the figure, with typical wavelengths and energies given against them. The various techniques which are employed in these different regions of the spectrum are also indicated in this figure, and to a very large extent the opening up of new regions of the spectrum has depended on the availability of new methods of producing radiation, or new techniques for its detection. However, for the pure scientist, the real interest lies in the different types of information which can be deduced by studying spectra in these different regions. Thus the initiation of a systematic study of physics can be linked back to the time of Newton, who was the first to demonstrate that a beam of white light could be separated into individual components, and hence the whole subject of spectroscopy was born. The detailed empirical study of spectral lines in the visible region during the latter part of the last century, culminating in the work of such scientists as Balmer, then provided the experimental background against which any successful theory on the nature of the atom could be tested. This challenge was, in due time, successfully met by the theories of Bohr, in the first place, and the more sophisticated treatment of wave mechanics which followed. This detailed study of spectral lines in the visible region thus enabled a comprehensive theory to be established on the nature of the outer sections of individual atoms, and in particular on the forces which hold the outer electrons to the nucleus, and the interactions between these electrons in their outer orbits.

The study of other radiations within the electromagnetic spectrum then extended away from the visible region in both directions to the higher energy, shorter wavelength ultraviolet radiation on the one hand, and to the longer wavelength, lower energy, infrared on the other. Analysis of the ultraviolet spectra, which could be studied experimentally with the advent of quartz lens systems and vacuum spectroscopy, then helped to confirm the picture of atomic structure which the visible spectra had already elucidated. The energy changes in this region were larger, however, and hence corresponded to electronic transitions across a wider energy range. The crucial relation which applies in any region of the spectra is, of course, the fundamental quantum condition, initially postulated by Bohr, that the frequency of the emitted radiation is given by

$$h\nu = E_1 - E_2 \qquad (1.1)$$

where E_1 and E_2 are energy levels of the initial and final states.

Initial work in the visible and ultraviolet region tended to concentrate on the lighter atoms, where the energies of the smaller number of electrons concerned could be more readily calculated. However, the same basic relationship governing the frequency of the emitted radiation will apply to the very much larger electron jumps that can take place in a heavy atom, when one of the inner electrons is removed, and an outer electron falls across the many intermediate levels to fill this vacancy. Such large energy changes produce the much more penetrating X-radiation as indicated in the next region of *Figure 1.1*. Although the techniques for the production of such radiation are very different from those employed in the visible and ultraviolet region, all of these three regions, which between them cover frequency and energy changes of about 10^5, arise from

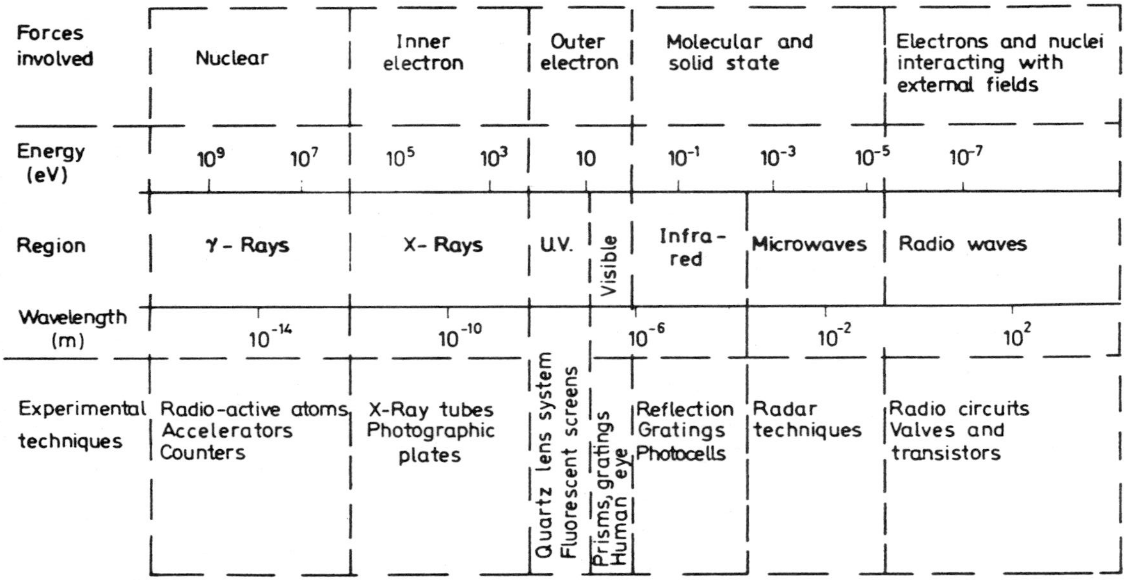

Forces involved	Nuclear	Inner electron	Outer electron	Molecular and solid state	Electrons and nuclei interacting with external fields
Energy (eV)	10^9　10^7	10^5　10^3	10	10^{-1}　10^{-3}	10^{-5}　10^{-7}
Region	γ - Rays	X- Rays	U.V.　Visible	Infra-red　Microwaves	Microwaves　Radio waves
Wavelength (m)	10^{-14}	10^{-10}	10^{-6}	10^{-2}	10^2
Experimental techniques	Radio-active atoms Accelerators Counters	X-Ray tubes Photographic plates	Quartz lens system Fluorescent screens / Prisms, gratings Human eye	Reflection Gratings Photocells / Radar techniques	Radio circuits Valves and transistors

Figure 1.1. The electromagnetic spectrum. Summary of the different types of radiation; their wavelengths, energies, methods used to produce and detect them, and physical forces associated with them

essentially the same physical interactions, i.e. from the forces which bind the electrons of an atom to its nucleus.

Since the energy transitions which give rise to X-rays are associated with electron jumps right across all the available orbits of heavy atoms, it might be supposed that the maximum energy changes had now been reached, and no radiation of higher frequency or energy could therefore exist. However, the initial experimental discoveries of radioactivity followed very quickly after the work on X-rays, and it was soon established that the γ-rays, which were emitted from some radioactive atoms, had frequencies and energies of even higher values than those associated with the X-rays. It followed that these could only originate from energy changes which were of an order of magnitude higher than those associated with the electrons in the atom, and hence such γ-rays must be produced from within the nucleus itself. Thus it was postulated that the nucleus could be excited to a higher energy state and would then return to its original condition, emitting a γ-ray in the process. In the same way that the details of the electronic energy levels, and forces which give rise to them, can be studied by analysing spectra in the visible, ultraviolet and X-ray regions, so the energy levels within the nucleus itself and the forces acting within it, can be studied by analysing the spectra of γ-rays emitted from them. Hence a new type of spectroscopy came into being some three decades ago, and has been supplying the detailed information on the nucleus, and its energy states, during recent years, in the same way that work on visible spectra supplied information on the electronic energy levels some seventy years previously.

This brief review of the different types of spectroscopy which exist across the whole electromagnetic spectrum can be completed by returning to the longer wavelength, lower energy, radiations by retracing one's steps through the X-ray, ultraviolet and visible region to the infrared region. Here the energy changes involved will be smaller than those associated with even single electron jumps in the outer levels of the atom. As a result weaker forces and interactions than those which bind the electron to the nucleus must therefore be responsible for radiation in this region, and such interactions are to be associated with those which bind atom to atom to form a molecule, rather than those within the individual atom itself. The study of spectra in the infrared region will therefore give very precise information on the molecular motions of molecules as a whole, such as the rotational states of diatomic molecules, or the vibration or bending associated with others. Analysis of the energies of these rotational or vibrational states can then be linked back directly to the study of the molecule itself. It is then possible to deduce structural parameters of

4

the molecule such as bond length and angles, together with information on the forces involved, from the observed absorption frequencies.

These general relationships between spectral line frequencies and molecular motion can then be traced further still into the longer wavelength, lower energy regions comprising the microwave, or radar, region. The energy changes involved are again becoming smaller and smaller, and the molecular motions now include the 'turning inside out', or inversion of such molecules as ammonia. The strengths of the forces concerned are also equivalent to those which act within a crystalline solid and hence analysis of spectra in this region has enabled very precise information to be obtained on solid state forces and interactions. Such results and their analysis do in fact form a main theme within other chapters of this book, and so they are not pursued in more detail here. In the same way later chapters explain in detail how information on the precise types of chemical binding within different chemical groups can be deduced from spectroscopy at radio frequencies, where the energy changes are smaller still, and only correspond to about one ten thousandth of an electron volt.

The main point of this introductory section, however, has been to serve as a reminder that all these different regions of the spectrum can be regarded as one integrated whole, and that spectroscopy can be conducted in any of these regions. The frequencies of the spectral lines then measured give precise information on the different types of forces and interactions, as summarised in *Figure 1.1*.

1. 2 Energy Levels and their Characterisation

It has been seen that spectroscopy can be carried out in all the different regions of the electromagnetic spectrum, and in each case the aim of spectroscopy is to determine a set of energy levels whether they are in molecules, atoms or nuclei. The transition which produces the observed spectral line is usually between one of the excited energy levels of the system being studied and its normal ground state. The different forces and interactions which produce and modify the energy level system will vary enormously according to the spectral region being investigated, but the idea of a ground state and the excited level remains the same, whether protons and neutrons inside the nucleus are being considered, or the interactions are those between different atoms in a polyatomic molecule.

In any such system there will be a certain configuration of the electrons, or nucleons, or other component particles, which have a minimum energy. For example, in the particular case of the inert

5

gases, the electrons line up their spins and orbital momentum to give a zero resultant which has the minimum amount of energy. A similar effect occurs in the magic number nuclei of nuclear theory. If the configuration of the molecule, atom or nucleus now changes from that of the ground state (e.g. an electron or neutron changes the orientation of its spin), then the new configuration will have an energy higher than that of the ground state. In moving up from the ground state to this excited level, the system will absorb radiation, the frequency of which is given by the quantum condition of equation 1.1, and similarly, if the system returns to the ground state from the excited level, radiation will be emitted with the same frequency. In any given system, whether it be the electrons around a single atom, the nucleons in the nucleus, or vibrations of a molecule, there will normally be many different possible excited states and hence a whole series of spectral lines can be observed from transitions between them. Some of these transitions may give rise to spectral lines in quite different regions of the electromagnetic spectrum (e.g. the blue colour from a crystal of copper sulphate is produced by transitions from excited states which have a different orbital momentum from the ground state, and which cause absorption in the optical region, whereas absorption lines can also be observed in the microwave region when a magnetic field is applied across the specimen—these being due to transitions between energy levels for which the orbital motion of the electrons is effectively quenched, and only the electron spin is changing). It is a very general feature of spectroscopy that measurements in different regions of the spectrum can often be compared and collated, and then the complete energy level system deduced from a combination of different experimental results.

The problem of the experimental spectroscopist is to measure the frequency of the emitted or absorbed radiation as accurately as possible, and this involves consideration of such questions as line-width, intensity and resolution. Once the experimental data has been obtained for as many transitions as possible in a given system, it is then the task of the theoretical physicist to produce a consistent model or theory. This must predict a comprehensive picture of the energy level system, so that the transitions between the different energy levels do produce the experimentally measured energy differences and frequencies. The first striking example of such a model was Bohr's theory of the hydrogen atom. In this case the experimental data consisted of the measured frequencies of spectral lines in the visible and ultraviolet regions, which had been summarised by experimentalists, such as Balmer, in the form of the empirical Balmer, Lyman and similar series. Bohr was then able to produce a model by applying Planck's quantum condition to the energy states of the Rutherford atom, which predicted a complete set of energy levels such that the differences between them accounted for the observed frequencies of the different spectral series.

Although this theory has been superseded by the more sophisticated treatment of wave mechanics, it does illustrate the general way in which progress takes place in spectroscopy, and the final step occurs when the theoretical explanation of the energy level system is used, in its turn, to give information concerning the forces and interactions which exist within the system being studied. In this way the experimental data which was obtained initially from the observed frequencies of spectral lines is used to check, modify and expand theories on atomic and molecular forces and interactions within the solid state.

In the above summary of the principles of spectroscopy, it has been assumed that all the energy levels have definite and precise values, and that the transitions, and spectral lines to which they give rise, have zero frequency spread. In practice, however, every form of spectral line has a finite width, and the reduction of the widths of the absorption or emission lines is one of the main problems which the experimental spectroscopist has to face. This width arises from an indefiniteness in the actual energy of the different energy levels, and can be caused by various factors such as thermal vibrations, collision with other molecules, or interactions with magnetic or electric fields surrounding the atoms or nuclei. Some of these factors can be reduced by altering the experimental conditions under which the observations are being made, and the reduction of linewidth to as small a value as possible is always desirable, since the main transition being investigated is often split by weaker interactions to give rise to a fine or hyperfine structure. Such fine structure can often only be resolved if the linewidth is made sufficiently narrow, and hence high resolution is normally required in any type of spectroscopy.

Even if all the external sources of broadening are removed and the atom is in a totally isolated state, there will still remain a definite spread in its energy levels which gives rise to the 'natural width' of the · observed spectral line. This spread in energy arises from the Uncertainty Principle which, in terms of an energy–time relationship, can be expressed as

$$\Delta E . \Delta t = h/2\pi \qquad (1.2)$$

This formulation of the Uncertainty Principle predicts that, if the measurement of energy can only be made over a limited time Δt,

7

then the observed value of the energy will be uncertain by an amount $\Delta E = h/2\pi.\Delta t$. For any given excited system, the maximum time for which such an observation could be made would be the lifetime of the excited state itself, and hence this lifetime will of necessity set a limit on the precision of the energy of that excited state. For example, if the lifetime of the excited electronic state of an atom was 10^{-8} s, then the energy spread of its associated level would be 10^{-26} J, and as a frequency spread this would correspond to a linewidth of 1.6×10^7 Hz, compared with an average frequency for a typical visible emission line of some 3×10^{14} Hz. There is therefore a fundamental limitation which prevents a resolution greater than one part in 10^7 in this particular case, and for even shorter excited lifetimes, the resolution would be correspondingly worse.

The actual value of the energy associated with the different levels in any atomic or molecular system can be specified in a variety of ways. The fundamental unit of energy is the joule, and the quantum of energy $h\nu$, associated with a radiation of frequency ν, can always be expressed in terms of joules with a value of h, the Planck constant, equal to 6.626×10^{-34} J/s. There are, however, other units than the joule which are often used in spectroscopy, and can be more convenient when comparing one set of energy levels with another. One of these is the electron volt, and which is in fact used in *Figure 1.1* to characterise the energies in different regions of the electromagnetic spectrum. This is the energy acquired by one electron in falling through a potential energy difference of 1 V. It can thus be very easily visualised as it measures the electron jump directly, and since the energy difference between the outer orbits of atoms are of the order of a few volts, it follows that any changes giving rise to radiation in the visible region will be of the order of a few electron volts, as is indicated in *Figure 1.1*. It should be realised, however, that the actual amount of energy associated with 1eV is extremely small, since it only measures the energy associated with one single electron, the actual conversion from joules to electron volts is given by 1 J = 6.24×10^{18} eV, or 1 eV = 1.60×10^{-19} J. Another parameter often used for characterising energy levels in spectroscopy, is the wave number, which is the number of wavelengths in the standard unit of length, 1m. This can be related to the other units by the general equation

$$\text{energy} = h\nu = hc/\lambda \tag{1.3}$$

and if the quantitative values of the Planck constant and the velocity of light are substituted, the value of the wave number of the energy level is given by

$$\text{wave number} = 1/\lambda = (\text{energy in electron volts}) \times 8.045 \times 10^5 \, \text{m}^{-1} \quad (1.4)$$

In a large number of publications in spectroscopy the wave number is still quoted in units of cm^{-1}, rather than m^{-1}, and the only point to remember in this connection is that since these are inverse units, the cm^{-1} is a hundred times larger than the m^{-1}. In this connection it is worth noting that energies can also be specified in terms of the equivalent temperature from the relation $h\nu = hc/\lambda = kT$. Substitution of the numerical value of the Boltzman constant, k, into this shows that the cm^{-1} is of the same order as the degree absolute. In fact 1 cm^{-1} = 1.436 K, and this enables very quick estimations to be made on the relative population of energy levels at any given temperature. Thus room temperature corresponds to an absolute temperature of some 300 K, or an equivalent energy level separation of some 250 cm^{-1}. It therefore follows that at room temperature excited energy levels higher than 250 cm^{-1} above the ground state will be very sparsely populated. In the same way it can be seen that for systems where the excited levels are relatively close to the ground state, such as ions in a crystalline lattice, the use of low temperatures may be necessary to produce a significant difference in population between the excited level and the ground state, and thus an observable absorption spectrum. This is quite an important point, especially in microwave spectroscopy, and is considered in much more detail in a later chapter.

1.3 General Methods of Spectroscopy

The methods employed to obtain and record spectral lines vary considerably from one region of the spectrum to another, but the basic principles of the experimental methods remain the same. Thus the essential requirements for the observation of absorption spectra at any wavelength are (i) a source of radiation, (ii) some form of absorption cell, or specimen holder, (iii) a detector with which to measure the intensity of the radiation after it has passed through the absorption cell.

When emission spectra are being studied, the source of radiation is replaced by an energy source, which excites the system under study to a higher energy level. This then returns to the ground state emitting a wavelength which can be detected in the same way as for the absorption spectra. In most regions of the spectrum it is

9

also necessary to have one further basic element in this system, and that is some means of selecting and isolating the different wavelengths which are being either absorbed or emitted. Thus in most cases the source of radiation is not monochromatic, and is emitting energy of a wide wavelength range. It will then be necessary to have some means of selecting one specific wavelength from this range; a device such as a prism is suitable.

The use of these general methods can be illustrated in the visible region by a typical spectrometer where the source of the radiation is usually some form of gas discharge, and after passing through the absorption cell, the different wavelengths are separated by means of a prism or diffraction grating. They can then be detected by direct visual observation, or recorded on a photographic plate. The principles of this technique extend with modifications into both the ultraviolet and infrared regions. In the former it becomes necessary to evacuate the system and use quartz prisms and windows to avoid continuous absorption by glass or air, while in the infrared region special detectors using quantum effects in the solid state often need to be employed.

In the high energy regions of the spectrum, individual wavelengths can often be produced by allowing accelerated particles to strike various targets, and induce transitions between the specific energy levels in the target material. Thus X-ray spectra can be produced by bombarding heavy atoms such as copper or tungsten with a beam of fast electrons. These will eject some of the inner electrons from the orbits of the copper atom, and allow an outer electron to fall into this level, emitting a high energy X-ray quantum in the process. Similarly, the higher energy γ-rays can be produced by bombarding nuclei with very high energy protons, to produce excited nuclear levels, which will emit γ-ray spectra as the nucleus returns to the ground state. In this wavelength region, the methods of detection consist of various types of solid state counters instead of the photographic plates, which are normally employed throughout the X-ray, ultraviolet, visible and infrared regions.

At the other end of the spectrum, that is in the longer wavelength microwave and radio regions, it is more usual to study absorption spectra. This is because the very small value of the quantum at these wavelengths makes it difficult to detect emitted quanta above the general background radiation. The advent of the maser and laser in recent years has somewhat altered this situation, and very high resolution emission spectra can now be obtained fairly readily in these regions, as will be discussed at some length later. There is one striking difference between this region and most of the others, however, in that there is no need for a dispersive element, such as a

10

prism or diffraction grating, since it is possible to produce highly monochromatic radiation from the electronic sources which are available in these regions. The radiation source, the absorption cell and detector are thus normally the only three essential elements required in spectrometers in these two regions of the spectrum.

It is interesting to note in this connection, that the advance of spectroscopy has been dependent to a large measure on the discovery of sources and detectors for radiation at new wavelengths. The ultraviolet and infrared regions were gradually opened up as new detectors were produced, and the extension of spectroscopy over the last two decades to the very high and very low frequency ends of the spectrum has been mainly due to the new sources of radiation that have become available in these regions. On the one hand very high energy accelerators have been built to allow excitation of highly excited nuclear states, and on the other hand all the war-time research on radar led to very stable and reliable microwave oscillators, together with a large number of other microwave techniques which were developed at the same time.

1.4 The Advent of Microwave Physics and Techniques

As can be seen from *Figure 1.1,* the expression 'the microwave region' refers to radiation of longer wavelength than the infrared, but shorter wavelength than that in the radio region of the spectrum. The term 'microwave radiation' is normally taken to refer to wavelengths between about 1 mm and 30 cm, although this is, of course, an arbitrary choice. Experimentally, however, it is a very significant region, since for some considerable time there was a gap in the complete electromagnetic spectrum at this point, and before 1940 there were very few sources of radiation, or methods of detection, available at these wavelengths. Thus on the one hand the optical techniques had been extended into the infrared region, whereas on the other hand development of the early work on radio waves had extended these to fairly high frequencies. The transit time effects of electron beams moving across radio valves had, however, set a limit for effective production of radio waves at frequencies of about 500 MHz, or 1 m wavelength.

The gap between the far infrared and the radio region was, in fact, only filled as a result of all the very active war-time research on radar. The new ideas and techniques of velocity modulation were developed at that time to produce valves like the klystron and the magnetron, which were no longer limited by the transit time effects of ordinary radio valves, and which were able to produce highly monochromatic, high power, radiation in this microwave region. At the same time all the techniques associated with waveguide propagation and waveguide components were developed, so that microwave circuits could be fabricated. These followed the same basic principles that had become well established in the lower frequency radio region, but the radiation now travelled inside a metal guide system, very much like transmission of light at much higher frequencies. One of the very interesting points about the progress of microwave physics has been the way in which the pure scientist and applied engineer have interacted so closely, to produce ideas and devices which have been of immense mutual benefit.

The normal picture of scientific progress involves an idea germinating in a pure science laboratory in a university, or government research station, where it is tried out in a somewhat crude form to check that the basic principles are in fact correct. This idea is then passed on to a pilot stage of development, and then into the main development laboratories of an industrial concern. The final step in this progress is the commercial production of the device for public consumption, and in this way there is a steady one-way flow of ideas and development from the pure scientist, to the applied scientist, to the engineer and production manager. However, in the field of microwave physics exactly the opposite has occurred, since the microwave region could not have been opened up effectively without all the initial work on radar by the applied engineers. Thus, in this case, the applied engineers were able to offer the pure scientist a whole series of tools and techniques ready made, with which the new region of the spectrum could be investigated.

This interchange has continued and has also often operated in the reverse direction since some of the early studies of the pure scientist in this region of the spectrum came up unexpectedly with very useful devices for the applied engineer. Thus the early work on microwave gaseous spectroscopy led to the development of the first 'maser' systems, in which ammonia molecules were employed, and which, in turn, produced the first atomic clocks. These entirely new references for the measurement of frequency and time have now taken over from the earlier standards linked with astronomical measurements, and some of the most important developments in modern industry, such as that linked with communication satellites, now rely on this particular application. In another field, the studies of the university research scientists on the energy levels of paramagnetic salts led to the development of the solid state 'maser' and 'laser' systems which now have immense practical applications in quite a variety of industries. Hence the wheel has turned full circle with the original techniques of the applied engineers enabling the pure scientist to open up an entirely new region of the spectrum, and then, in turn, applied devices have emerged from these new studies which have been of immense benefit to the applied engineer. This interaction and interchange in fact continues to develop, since one of the more recent applications of the laser systems has been in the field of pure research to very high resolution spectroscopy, enabling the pure scientists to obtain much more detailed information about the nuclear hyperfine interactions within the systems they are studying.

It will be clear from the figures quoted at the end of the last section that the frequencies and wavelengths of the microwave region, which correspond to a wave number of about 1 cm^{-1}, also correspond to energy changes of about one ten-thousandth of an electron volt, and hence are very much smaller than those associated with normal atomic transitions which give rise to optical spectra. The microwave transitions are, in fact, produced by the weaker forces of a molecular, or solid state, character which bind the different atoms together to form molecules, or hold them within a crystalline lattice. The study of spectra in this region will therefore give detailed information concerning these interatomic binding forces, and also on the magnitude and symmetry of the internal fields or solid state interactions. However, as is often the case in other forms of spectroscopy, large amounts of information also become available from the splittings which are observed in the initial spectral lines. These splittings reflect second-order interactions in the energy levels concerned, and can sometimes give very precise additional information on interactions with nuclear spins, or other internal effects within the system. One of the great advantages of working in the microwave, or radio, region is the very high resolution available, due to the very monochromatic radiation which can be produced, and the high stability of the measuring and detecting equipment. Thus if molecular rotations are studied from their microwave spectra, much more detailed and exact information can be obtained from them than from the corresponding spectra observed in the infrared region, where the resolution is much lower, and the detecting methods less efficient.

It is not appropriate in an introductory book of this nature to present a detailed treatment of the theory of microwave production and propagation, but a brief summary of the way in which this radiation is transmitted through waveguides is necessary in order to appreciate the main features of spectroscopes working in this region.

As mentioned earlier, the propagation of microwave radiation down a waveguide is much more similar to the transmission of light down a hollow tube than to the transmission of alternating current

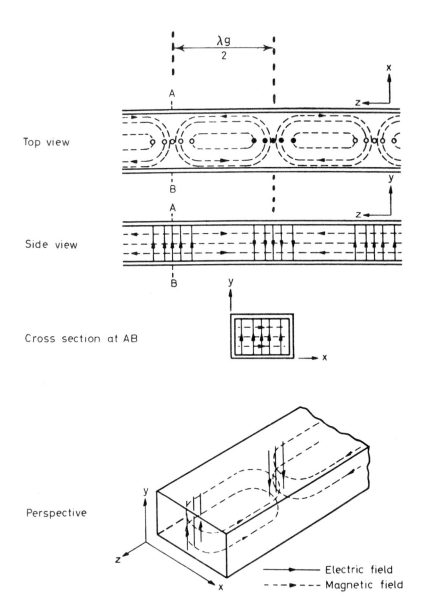

Top view

Side view

Cross section at AB

Perspective

Electric field

Magnetic field

Figure 1.2. Electric and magnetic fields for dominant waveguide mode. The fields shown are those which exist for a travelling wave, with the maxima of the electric and magnetic fields moving along together

14

down two conducting wires. In considering the properties of microwave radiation, we are essentially dealing with the motion of the electric and magnetic fields which make up the radiation front, rather than with the voltages and currents flowing in the surrounding walls. These surrounding walls do serve the very useful purpose, however, of defining precisely in which direction the electric and magnetic fields are oriented. Although some complicated patterns of electric and magnetic fields can be set up within a waveguide system, the waveguides are normally designed so that only one simple field pattern can exist within them, and this is known as the 'dominant mode'. For a simple rectangular waveguide, with the longer side of its cross-section equal to 'a' the condition for transmission of this dominant mode alone is given by $0.706. \lambda > a > 0.50. \lambda$. Thus if a wavelength of 3 cm is to be propagated, the waveguide size should be chosen so that its broader dimension is greater than 1.5 cm and less than 2.1 cm. In practice the guide is chosen so that it is well away from the cut-off condition, given by $\lambda = a/2$, since high attenuation of the microwaves will occur near this condition, owing to the longer paths for current flow in the waveguide wall.

The way in which the electromagnetic fields are oriented for this dominant mode in a rectangular waveguide is shown in *Figure 1.2*. The lower part of this figure gives the perspective view, while the upper portion gives views from the top and the side. It can be seen that the electric vectors are in the form of simple straight lines stretching from the centre of the two broad faces of the guide, and repeating in opposite directions at distances of half a guide wavelength. The magnetic lines of force form closed loops, however, in the way indicated in the figure. For a travelling wave the maximum value of the magnetic field will occur at the same place as the maximum value of the electric field, as shown in the figure, so that the Poynting vector, which determines the energy flow down the guide, and is given by the vector product of these two field values, is at a maximum.

It is important to know precisely where the maxima of these field patterns are, since these will be the oscillating fields that couple to the electrons or nuclei being studied, and it will therefore be necessary to place the specimens being investigated into these regions of maximum field. Moreover, an understanding of the distribution of these field patterns also makes it clear how they can be launched down the waveguide system, either from a stub aerial which will initiate the electric vector, as shown in *Figure 1.3 (a)*, or by a continuous shorting loop at the end of the coaxial line, which will set up the magnetic field as indicated in *Figure 1.3 (b)*. The operation of microwave attenuators, in which a piece of card coated with lossy material such as graphite,

(a)

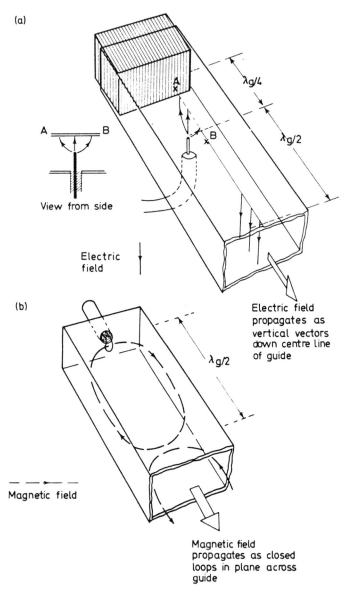

A ——————— B

View from side

Electric
field

Electric field
propagates as
vertical vectors
down centre line
of guide

(b)

λg/2

Magnetic field

Magnetic field
propagates as closed
loops in plane across
guide

Figure 1.3. Methods of launching electromagnetic wave into wave-guide: (a) electric field excitation from stub aerial; (b) magnetic field excitation from loop

16

can be lowered through the centre of the broad side of the guide into the region of maximum electric field, can also be understood from this field pattern consideration, as indicated in *Figure 1.4*. No

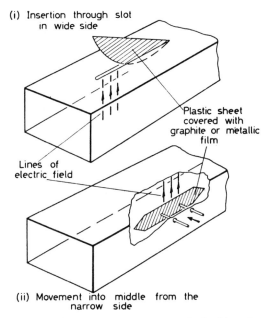

(i) Insertion through slot in wide side

Plastic sheet covered with graphite or metallic film

Lines of electric field

(ii) Movement into middle from the narrow side

Figure 1.4. Microwave attenuators. Gradual insertion of the lossy material into the region of electric field strength dissipates energy from the wave

detailed description, or consideration of these various microwave components will be pursued in this book, but it is hoped that these general references to the distribution of the oscillating electric and magnetic fields down the waveguide system will be sufficient to give a clear idea of how the microwave energy is being transported, and how it can interact with the components of the microwave spectrometer system, and with the specimens which are inserted into it.

1.5 Early Work in the Microwave Region

The very first experiments on gaseous spectroscopy in the microwave region were carried out in 1933 by Cleeton and Williams, who showed very considerable skill in obtaining any results at all with the techniques then available. They investigated the absorption of ammonia vapour around a wavelength of 1.5 cm, since theoretical predictions had shown that the inversion absorption frequency for the two positions of a nitrogen atom should be in this region. The ammonia molecule is in the form of a pyramidal structure with the nitrogen atom above the plane formed by the three hydrogen atoms, as indicated in *Figure 1.5*. It is possible for the nitrogen atom to be in equilibrium on either side of the plane formed by the hydrogens, however, and at any normal temperatures it will, in fact, oscillate between these two possible positions, the molecule inverting, or turning inside out, in the process. Calculations had suggested that the frequency at which these oscillations take place should be at around 25 000 MHz, and therefore at a wavelength of somewhat over 1 cm. The apparatus which Cleeton and Williams used was very direct and simple, the ammonia gas being contained in a rubberised cloth bag at atmospheric pressure and placed in a beam formed by focusing the microwave radiation with parabolic mirrors. They produced the radiation itself from very small split-anode magnetrons with anode radii of less than ½ mm, these being the earliest forerunners of radar valves which were developed some years later. The microwave powers available, and their detection methods, resulted in a rather low sensitivity for the experiment,

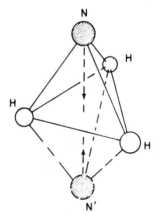

Figure 1.5. Structure of ammonia molecule

but they were able to show that there was a broad absorption band centred on a wavelength of about 1.25 cm, the relatively high pressure of the gas preventing any further resolution of the line.

Apart from this one experiment, spectroscopy in the microwave region lay dormant until after the war, when the enormous advance in microwave techniques opened up this new field of research. Thus the invention and development of the klystron and magnetron for

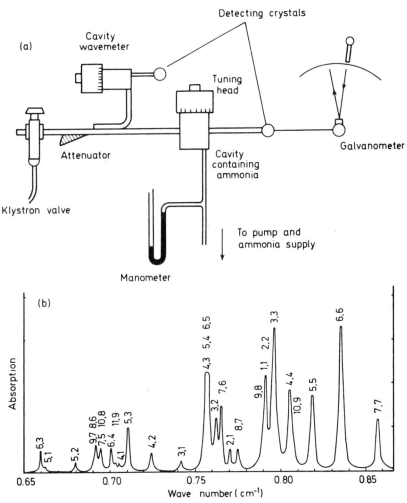

Figure 1.6. Initial study of NH₃ by microwave spectroscopy: (a) block diagram of original spectroscope; (b) spectrum observed around 1.25 cm wavelength

radar purposes had produced highly robust oscillators which could generate radiation of high power, and the frequency of which could be accurately controlled and measured to one part in 10^7.

19

Thus it was not surprising that the work on spectroscopy in the microwave region started at several different laboratories as soon as the war was over, and ammonia gas was again the first molecule to be studied. Bleaney and Penrose who were working in the Clarendon Laboratory at Oxford were the first to publish their results, and a block diagram of their apparatus, together with the first spectrum observed, is shown in *Figure 1.6*. The radiation is produced by the klystron valve and then passed down the waveguide run to the cavity resonator, which acts as the absorption cell. A simple semi-conducting rectifying crystal was employed as a detector, mounted across the waveguide on the output side of the cavity. The crystal is coupled directly to a sensitive galvanometer, so that the level of the microwave power passing through the cavity is determined directly from the reading on the galvanometer scale. A directional coupler led off from the main waveguide run to a second cavity resonator, which acted as a wavemeter, and from which the wavelength of the microwave radiation could be accurately measured.

The reading on the galvonometer scale was then noted for each different setting of the klystron—the wavelength being measured for each reading and the cavity resonator always being kept tuned to the klystron frequency. In this way a plot of microwave absorption against frequency was obtained and the series of spectral lines was identified, as shown in *Figure 1.6 (b)*. The resolution shown in this spectrum was in fact observed only at low pressures of the ammonia gas, since collision broadening between the gas molecules will occur at the higher pressures and only the single broad line originally observed by Cleeton and Williams will be obtained.

The existence of a large number of spectral lines in these results may be somewhat unexpected, since our earlier consideration had suggested that the ammonia molecule would just invert at one single frequency. However, this spectrum is a very good illustration of the fact that second order additional interactions can often affect the main transition which is being observed, and produce fine splittings in it. In this particular case it is the rotational motion which the ammonia molecule possesses, at the same time as its inversion, that produces the splittings of the main line. Thus if the ammonia molecule is rotating about an axis normal to the plane of the hydrogens, the centrifugal force acting on the three hydrogens can be visualised as tending to move these further apart, and thus reduce the potential barrier through which the nitrogen atom has to move, and hence also alter the frequency of inversion. This type of rotation determines one set of quantum numbers shown against the spectral lines in the figure, while the other is determined by rotation in the

20

other plane, of an end-over-end type, which will also produce a second-order shift in the frequency of the main inversion. It becomes clear immediately that study of such spectra gives very much more information than just the single energy difference corresponding to the inversion between the two positions of the nitrogen atom, and, in fact, detailed values on the bond length and angles for the molecule as a whole can be deduced from them.

Similar work was also being carried out in America at the same time by Good, but he was using a long waveguide type of absorption cell, in place of a cavity resonator. As it turned out this was experimentally a more effective technique, since it avoided the saturation broadening which the concentration of microwave power in the cavity produced in Bleaney and Penrose's equipment. This is a point which is discussed in more detail later, but the net effect was that Good was able to obtain higher resolution in his measurements and was able to resolve out the hyperfine structure due to the interaction with the nitrogen nucleus. This discovery of the hyperfine structure associated with the spectra from gaseous molecules aroused immense interest among physicists and chemists. Thus it became possible not only to evaluate the interatomic distances and molecular structure from the central frequencies at which the spectra were observed, but also to measure nuclear properties from the hyperfine splittings that were obtained. In particular, the nuclear spins of many nuclei could be determined, together with the electric quadrupole moments of the nuclei, which is a measure of the distortion of the nucleus from spherical symmetry. The way in which this wealth of information became available from the high resolution microwave gaseous spectroscopy that then developed is discussed in detail in the next chapter. Here it might just be noticed that these initial measurements served to underline one of the great advantages of the microwave region, i.e. the very high resolution that is potentially available in measurements at these frequencies.

At the same time as different laboratories were taking up the study of gaseous spectra at microwave frequencies, others were attempting to apply the new techniques available in the microwave region to studies of the solid state. In particular Zavoisky in the U.S.S.R., and Cummerow and Halliday in America were attempting to detect the difference in energy which unpaired electrons would have when placed in a strong external magnetic field. This would align them either with, or against, the field and the two energy states might be detected by resonance absorption of incoming electromagnetic radiation. The vast majority of electrons in any atomic or molecular system are in fact paired off with others of equal and

21

opposite spin, and all the main types of chemical bonding use this pairing to produce a lower energy level and a strong bond. (Thus in an ionic bond such as sodium chloride, one electron is actually transferred from the sodium to the chlorine to leave an even number on the sodium and make an even number of electrons on the chlorine; in the case of covalent bonding between similar atoms, such as a carbon–carbon bond, one electron is taken from each atom and shared to produce an electron pair bond, the wave function of which embraces both atoms.) There are, however, some exceptions to this rule such as the transition group atoms, where a lower shell of electrons, such as the $3d$ orbits, remains unfilled while an outer layer, such as the $4s$, fills up and takes part in the chemical bonding. Other examples include the electrons associated with defects and damage centres in solids, and the active free radical centres which are often found in organic molecules. In such cases as these, there will be an unpaired electron in the atom or molecule, which therefore has both a spin and an associated magnetic moment, which is not compensated by any other of the electrons present. The atom or molecule can therefore be regarded as possessing a small permanent magnet, or dipole moment, and this magnetic moment will now align itself in any applied external magnetic field.

The spin quantum number of an individual electron is equal to ½, and since quantum numbers can only differ by unity, it follows that there are only two possible orientations for a resolved component of the electron spin in an applied magnetic field, i.e. either along the field with the quantum number Ms = + ½ or lined up against the field with a quantum number Ms = −½. It therefore follows that if a substance containing such unpaired electrons is placed in the external magnetic field, the electrons will be sorted into two groups, i.e. those lined up with, and those lined up against, the field. Moreover, these two groups will be separated in energy, since those lined up with the field have their energy lowered, while those lined up against the field are raised by a corresponding amount. The actual magnitude of the energy change in each case is given by the value of the moment resolved along the field multiplied by the strength of the field itself. The magnitude of the resolved magnetic moment will be given by ½$g\beta$, where ½ is the spin quantum number, β is the Bohr magneton, which effectively converts the units of angular momentum associated with spin to units of magnetic moment; while g is a parameter which measures the contributions of the spin and the orbital motion of the electron to its resultant magnetism, and also the effects which its environment might have on either of these. For a completely free electron g would have a value of just on 2.0, but its value in the solid state can depart quite noticeably from this, and determinations of its magnitude and variation have proved to be one of the most useful features of this type of spectroscopy.

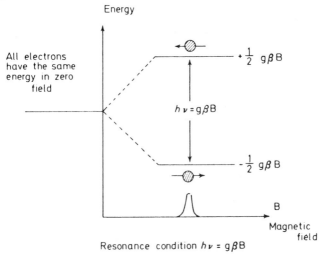

Figure 1.7. Basic principle of electron spin resonance. For a free electron the resonance frequency is given by hν = 2.8 x 10^{10} B Hz where the magnetic field, B, is in teslas

The net effect of the applied magnetic field on the unpaired electrons is illustrated in *Figure 1.7.* In the absence of any applied magnetic field, the electrons will all have the same energy, whatever direction their spins and magnetic moments are pointing, but when the field is applied, they separate into the two groups as shown, and it is clear that there is now a specific energy gap between them, equal to $g.\beta.B$. It follows, therefore, that if electromagnetic radiation of frequency ν, such that

$$h\nu = g.\ \beta.\ B \qquad (1.5)$$

is fed to the molecules, absorption of this radiation should occur as electrons are excited from the lower to the higher state. If the magnitude of the energy gap and the associated resonant frequency are calculated for the largest kinds of magnetic field that can readily be obtained with an electromagnet (i.e. about 1 tesla), then it is found that the frequency required for such resonance absorption is about 28 000 MHz. It is thus well into the microwave region of the spectrum,

23

and fairly close to the resonance frequency for the ammonia molecule discussed in the last section. It is therefore apparent that, just as the ammonia molecule can absorb microwave radiation when this is equal to its natural inversion frequency, so unpaired electrons in a solid can also absorb microwave radiation when the splitting of their two energy states matches the frequency of the incoming microwaves. This is the principle of the second branch of microwave spectroscopy that followed out of all the war-time work on radar, and came to be known, first as paramagnetic resonance, and then, more generally, as electron spin resonance.

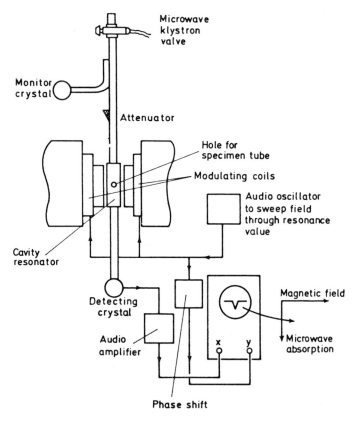

Figure 1.8. Simple electron resonance spectrometer with video display

The basic experimental equipment will be very similar to that used for the gaseous studies, as shown in the *Figure 1.6,* but a large applied magnetic field must now be added, and the basic features of a simple electron resonance spectrometer are then as shown in *Figure 1.8*. It will be seen that this has one other additional feature in the form of an a.c. method of displaying the absorption line on a cathode ray oscilloscope, instead of a point by point tracing of it from the d.c. readings of a galvanometer. This can be readily achieved in the case of the magnetic resonance experiments by adding a small a.c. modulation on to the large d.c. value of the magnetic field. This small modulation will then sweep the field back and forth through the resonance condition, and if the time base of the oscilloscope is kept in synchronism with the magnetic field modulation, the absorption line can be directly traced out on the oscilloscope screen as shown.

Fairly soon after the initial measurements on electron resonance, it was discovered that hyperfine structure could also be observed in these spectra, again providing a very sensitive and sophisticated tool for probing the internal interactions within a solid. It would probably be fair to say for both electron resonance and microwave gaseous spectroscopy, that the full realisation of the great potentiality of these new types of spectroscopy came with the discovery of the ease with which high resolution hyperfine splittings could be detected. The particular applications of electron resonance at microwave frequencies, together with more details of the techniques that are employed are discussed at length in Chapter 3, following the further treatment of gaseous spectroscopy.

1.6 Initial Studies on Radio-frequency Spectroscopy

The term 'nuclear magnetic resonance' is applied to experiments which are identical to those just described for electron spin resonance, with the exception that the electrons are now replaced by nuclei and the magnetic moments which are being orientated in the applied magnetic field are those of the individual nuclei, rather than those of the unpaired electrons. The basic principle of nuclear magnetic resonance can therefore also be represented by *Figure 1.7*, but with the magnetic moments of the electrons replaced by those of nuclei, such as protons. The resultant energy gap is then $\frac{1}{2} g_N \beta_N B$, where β_N, the nuclear magneton, is some 2000 times smaller than the Bohr magneton, since the expression for the magneton ($eh/2mc$) has the mass of the particle in the denominator rather than the numerator. Hence the energy level separation for nuclear moments in the same magnetic field will be some two thousand times smaller than for the electron case. Nuclear resonance investigations of protons will therefore require frequencies of the order of 15 MHz, and hence in the radio frequency, rather than the microwave, region.

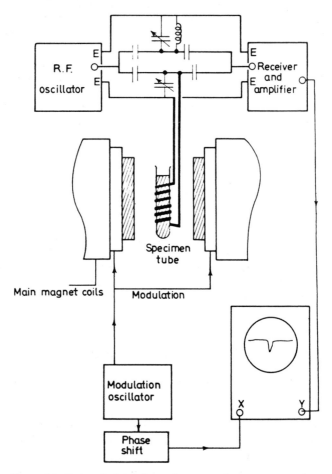

Figure 1.9. Basic principle of a nuclear magnetic resonance spectro-meter. This spectrometer employs a balanced r.f. bridge. Another system, employing nuclear induction is shown in Figure 4.1

Such work on nuclear magnetic resonance was being undertaken at almost the same time as the initial studies on electron resonance. The techniques are now very different, of course, since lumped circuit constants, consisting of condensers and inductances, are employed instead of the cavities of the microwave region, but the same basic principles underlie such a radio frequency spectrometer, and are illustrated in *Figure 1.9*. In the early studies of nuclear magnetic resonance two independent groups made the first important contributions, and the first of these to report their observations were Purcell, Torrey and Pound, who used a radio frequency bridge system, as represented diagrammatically in the figure. The extra absorption produced in the sample, when resonance takes place, unbalances the arm of the bridge in which its inductance is situated, and hence an off-balance signal is sent to a detecting system and can be displayed on the oscilloscope by the use of a small modulation, as described for *Figure 1.8*.

The other group who initiated work in this field consisted of Bloch and his colleagues in California. They employed a somewhat different technique of nuclear induction, which has since come to be adopted as the more normal method of studying spectra at these frequencies, and is described in much more detail in Chapter 4.

However, it should be mentioned that some years before these initial experiments on radio frequency spectroscopy of the solid state, the basic principle of radio frequency resonance had already been applied in the study of atomic and molecular beams. Atomic beams were first used in the classic experiment of Gerlach and Stern, who fired a beam of silver atoms in a path down an inhomogeneous magnetic field and were able to show from the separation of the beam into two distinct traces that the electron must have a quantised spin and magnetic moment. The next step was taken by Frisch and Stern, who used a beam of molecules instead of atoms, and were thus able to detect the effect of a field gradient on the nuclear magnetic moments. Thus if molecules are used, the electron spin and magnetic moments cancel, and the only resultant moment left for the whole molecule is that due to the nuclei. However, the deflection in this case is extremely small compared with the atomic beams, because of the very much smaller value of the nuclear magnetic moment. Then in 1938 Rabi and his colleagues introduced an entirely new technique which produced a very great increase in accuracy and sensitivity, and more or less rendered all the older methods obsolete. In addition to passing the beam down the long inhomogeneous magnetic field, they inserted a small length of homogeneous d.c. magnetic field, across which an oscillating radio frequency field was also applied, as illustrated in *Figure 1.10*. This was placed in the centre of the beam path between the two long deflecting magnets of opposing field gradients. If the strength of the magnetic field in the central region and the frequency of the

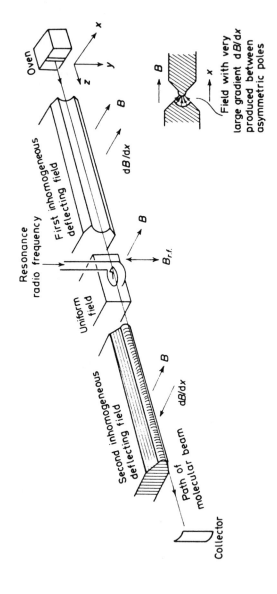

Figure 1.10. Radio frequency resonance methods as applied to atomic and molecular beams. The deflection in the Ox direction produced by the first inhomogeneous field is exactly cancelled by the second, unless resonance re-orientation of the nuclear spin occurs in the central region

current in the radio frequency coil are adjusted to fulfil the resonance condition $h\nu = g_N \beta_N B$ for the nuclei traversing the equipment, then the nuclear spins will change their orientation as they pass through the central region, absorbing the radio frequency energy in the process. This resonance absorption can be detected from the fact that the second deflecting field will no longer re-focus them onto a collector, and there will thus be a sharp dip in the value of the detecting current when the resonance condition is fulfilled. This resonance method was very rapidly applied to determine the moments and spins of a large number of nuclei, and there is no doubt that this work helped to germinate the ideas which were to flourish in the field of radio frequency resonance of the solid state immediately after the war, when much more sensitive equipment was available.

In the following sections of the book, the different topics are treated in a more methodical way, without too much reference to the chronological order in which specific discoveries were made, but it was felt that a brief historical survey in this introductory chapter might help to give a suitable background, against which spectroscopy in these two new regions of the spectrum could be discussed.

Chapter Two

Microwave Spectroscopy of Gases

2.1 Comparisons and Contrast with Infrared Spectroscopy

It was seen in the last chapter that the energy changes associated with both the infrared and microwave regions of the spectrum were significantly smaller than those associated with the electronic transitions of the visible and ultraviolet regions. In fact, the interactions producing these energy changes are of a totally different nature, being the interactions and forces *between* atoms, which bind them together to form molecules, rather than the electron–nuclear attraction within an individual atom. Hence it follows that both infrared and microwave spectroscopy of the gaseous state will be measuring energy levels associated with molecular motion, and the analysis of these spectra should therefore give information on molecular force constants and molecular structure.

It is also clear, however, that there is a significant difference between the typical energy studied in the infrared region, compared with the energy of a typical microwave quanta, the latter being some thousand times smaller. Hence infrared spectroscopy will be concerned with the larger and often more complex energy changes associated with molecular motion, whereas microwave spectroscopy will be studying the small energy changes associated with simpler motional states. In a general way, these two can be separated into the higher-energy vibrational states of molecules, and the lower-energy rotational motion.

It is probably easier to visualise rotational motion in the first place, and if the simple case of a diatomic, or linear triatomic, molecule is visualised, then one type of rotational motion will be when the whole molecule spins end-over-end around an axis passing through the centre of gravity of the molecule as a whole, and normal to the line joining the atoms, as illustrated in *Figure 2.1.* The energy to be associated with such a rotating molecule can be related very simply to its angular momentum and its moment of inertia, and for all linear molecules this energy is given by the very straightforward expression

$$E_{rot:} = J.\,(J+1).\,h^2\,/8.\pi^2\,.\,I \qquad (2.1)$$

which can be written as

$$E_{rot:}/h = J\,(J+1).\,B \qquad (2.2)$$

where J is the rotational quantum number, I is the moment of inertia of the molecule, and B is called 'the rotational constant' and will be seen to have the units of frequency and hence can be expressed directly in MHz.

The more general case of non-linear molecules will possess three distinct moments of inertia, I_A I_B I_C, with associated rotational constants A,B,C respectively. However, a large group of molecules known as the symmetric top molecules have two moments of inertia which are in fact equal, and under these conditions it can be shown that the energy of rotation then becomes

$$E_{rot:}/h = J.\,(J+1).\,B + K^2\,.\,(C-B) \qquad (2.3)$$

where K is the quantum number associated with rotation about the molecular symmetry axis, such that the internal angular momentum about this axis is equal to $Kh/2\pi$.

It can be readily seen that for both linear and symmetric top molecules the frequency of the transition from the J rotational state to the $(J-1)$ state will be given by

$$h\nu = E_{rot:\,J} - E_{rot:\,(J-1)}$$
$$\text{therefore } \nu = J.(J+1).\,B-(J-1).\,J.\,B = 2.B.J \qquad (2.4)$$

and hence the actual frequency of absorption is equal to $2.B.J$. The observed spectrum in these cases should therefore consist of a set of evenly spaced lines with a separation equal to $2B$, as indeed is found to be the case, but with minor corrections due to centrifugal distortion and other second-order effects.

A molecule that has been studied in great detail in this connection is OCS, and a detailed comparison of the observed rotational transitions for this molecule show that $2B$ is equal to a little over 12 000 MHz.

31

This is therefore a molecule for which the lower transitions will lie in the microwave, and not in the infrared, region of the spectrum.

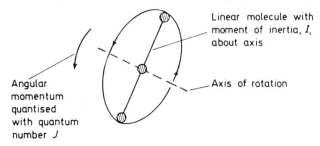

Linear molecule with moment of inertia, I, about axis

Axis of rotation

Angular momentum quantised with quantum number J

Figure 2.1. Rotational motion of linear molecule. Energy of rotation equals $J(J + 1)\ h^2/8\pi^2 I$ where J is the rotational quantum number, and I is the moment of inertia about the axis of rotation

It is clear from equation 2.4 that direct measurement of the spectral line frequencies will enable the constant B to be very readily determined, and since this is equal to $h/8\pi^2 I$, the moment of inertia of the molecule will also be determined at the same time. If the masses of the atoms are known accurately, as is normally the case, then it is a very simple matter to deduce the interactomic spacing between these atoms and hence the chemical bond length from their known masses and the measured moment of inertia of the molecule as a whole.

When vibrational motion is considered, however, the energy changes are significantly greater than those associated with change in rotational motion. To a good order of approximation, the different types of vibration which molecules will undergo can be considered as simple harmonic motion, a restoring force being provided by the interatomic binding forces present in the molecule. Distortion of the shape of the molecule by even small amounts will bring fairly strong restoring forces into play, and hence the frequencies and energy changes associated with the simple harmonic oscillations will be quite large. As a general expression, the energy associated with any vibrational motion, can be written as

$$E_{\text{vibrational}} = h\nu.\ (V + \tfrac{1}{2}) \qquad (2.5)$$

where ν is the actual frequency of the simple harmonic oscillations and V is the vibrational quantum number. It should be noted that even if the quantum number is zero, there is still a certain amount of vibrational energy left in the molecule, known as the zero point energy. This must be present otherwise the Uncertainty Principle would be denied because the position and momentum of the atoms in a molecule completely at rest could be precisely determined. The magnitudes of the vibrational frequency, v, will vary appreciably, depending on the type of motion and on the particular atoms contained within it, but typical stretching frequencies would be 10^8 MHz (about 3000 cm^{-1}) and hence around 1000 times greater than rotational frequencies in the microwave region.

The simplest type of vibrational motion that can be pictured in a molecule is the stretching which will take place when two atoms move slightly further apart along their common axis, and the restoring forces then pull them back past the equilibrium position and a simple harmonic oscillation thus results. This stretching motion is illustrated in *Figure 2.2 (a),* and other types of vibrational deformation are also

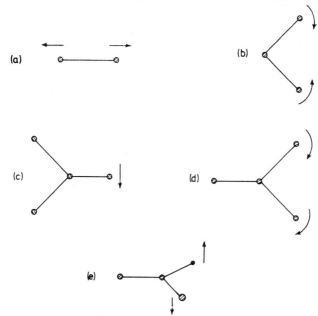

Figure 2.2. Different types of vibrational motion: (a) stretching; (b) angle bend or scissoring; (c) deformation; (d) rocking; (e) twisting

shown in that figure, including the increase and decrease of bond angle (b), a simple deformation (c), a rocking motion (d), and a twisting motion (e). These illustrations do not exhaust the possibilities, since it has already been seen that the ammonia molecule can undergo an inversion, or 'turning inside-out' type of vibration, and there are other more complex modes, such as 'ring-breathing', when the whole of the carbon ring system gradually expands and contracts, together with other simpler variations such as 'wagging', and 'scissor motion', which two hydrogen atoms attached to a carbon atom can undergo. It is not appropriate in this text, however, to consider the details of all such molecular motions, but to make the general point that these distortions involve fairly strong forces, and in nearly every case the sets of absorption frequencies associated with them are to be found well in the infrared region of the spectrum and not in the lower energy microwave range.

The other general comparison that can be made between infrared spectroscopy and that in the microwave region, is between the different types of experimental equipment that are employed. A simple gaseous microwave spectroscope has already been illustrated in *Figure 1.6,* and from the general comments also made in Chapter 1, it will be evident that the essential difference between this and an infrared spectroscope will be the necessity of some kind of dispersion element in the infrared spectrometer. Thus, apart from recent work using laser sources, most infrared spectrometers use a broad-band source of radiation, normally in the form of a heated element covered with material which will produce a continuous range of emitted wavelengths. If absorption of individual wavelengths is to be studied, it is then imperative that a selection is made from this broadband background radiation, which must therefore be passed through either a prism or diffraction grating before reaching the detecting equipment. The other general point to note is that in most infrared spectroscopes mirrors, rather than lenses, are employed for focusing the radiation, because of the absorption which will occur in most types of lens material. The prisms themselves have to be fabricated from carefully selected material, the nature of which may change according to the wavelength range being studied. Thus fused silica will be satisfactory down to frequencies of 2800cm^{-1} (about 10^8 MHz), but caesium bromide or iodide would be necessary for frequencies below 200 cm^{-1} (6×10^6 MHz). Infrared spectrometers therefore represent a special case of the more general type of spectrometer developed in the optical region, and, compared with microwave spectrometers, have the two serious disadvantages of (i) broad-band sources which therefore cannot concentrate the power into a single wavelength, and (ii) transmission systems which have to employ carefully selected material for efficient operation of the spectrometer.

2.2 Early Studies on Microwave Gaseous Spectroscopy

The very earliest measurements on microwave gaseous spectroscopy have already been considered in section 1.5 and a brief description of the first spectroscope used by Bleaney and Penrose was summarised diagrammatically in *Figure 1.6*. A large number of technical improvements have taken place in the design and construction of microwave spectroscopes since that time, but it is not appropriate to summarise these in detail in a book of this nature. However, it is important to describe one particular technique that was introduced because it has far-reaching implications for the theoretical analysis of the observed spectra in addition to the increased sensitivity which it gives to the spectroscope itself.

This technique is known as 'Stark modulation' and depends on the fact that a d.c. electric field applied across the gas in a microwave absorption cell will produce a splitting of its energy levels and thus also of the observed spectral lines. Both the Stark effect and the Zeeman effect have, of course, been observed earlier in atomic spectra in the visible region. They both arise from the different energies which are associated with the different orientations that the vector, J, representing the total angular momentum of the molecule, can take up with respect to a direction fixed in space. Thus if a molecule with a total angular momentum quantum number J is considered there will always be $(2J + 1)$ ways in which this total momentum can orientate in space, such that its projection along a fixed direction in space has resolved quantum numbers differing by 1, as illustrated in *Figure 2.3a*.

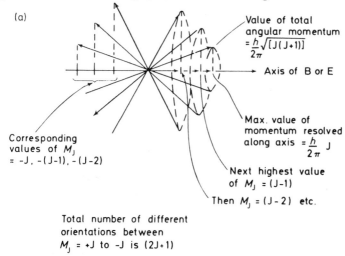

Figure 2.3. Zeeman and Stark effects: (a) The $(2J + J)$ orientations which a molecule can take up in an applied field are shown

35

Zeeman effect of atomic sodium

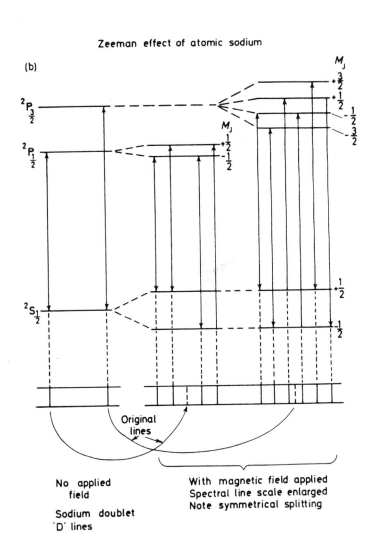

Figure 2.3. (b) A magnetic field produces linear energy differences between these, and a symmetrical 'Zeeman effect' is observed

Stark effect of atomic sodium

(c)

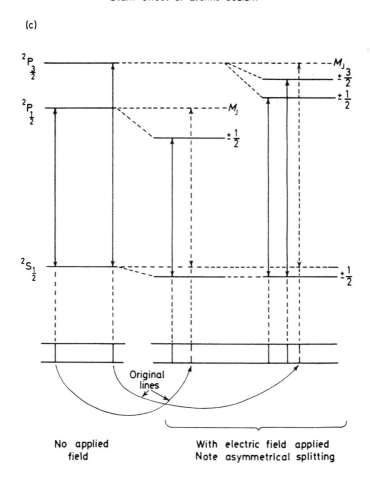

Figure 2.3. (c) An electric field produces energy differences which vary with the square of the field and an asymmetric 'Stark effect' is observed

In the absence of any external electric or magnetic field these $(2J + 1)$ states are all equivalent and have the same energy. However, when an external electric or magnetic field is applied a definite axis exists along which quantisation can take place, and the different resolved components along this axis of quantisation will now have different energy values for their corresponding orientations. Each of the $(2J + 1)$ sublevels, corresponding to a different M_J quantum value, now have a slightly different energy and this new series of energy levels will produce a splitting of the previously observed single absorption line into several different components. If the interaction is of a magnetic nature and occurs between an applied magnetic field and a permanent magnetic dipole moment of a molecule, as occurs in electron resonance, then there will be a linear relation between the energy shift and the applied magnetic field and symetrical splitting of the line will be produced. An applied electric field, however, does not act on the magnetic dipole moment of the molecule but on its electric dipole moment. In the case of atomic spectra the individual atoms do not usually possess a permanent electric dipole moment but one can be induced by the application of the electric field itself. In this case the splitting produced by the applied field will vary as the square of the magnitude of the field since it will be dependent on both the magnitude of the field itself and on the magnitude of the induced dipole moment which is itself field-dependent. It will also follow from this that the energy levels corresponding to the $\pm M_J$ components will have the same energy, and hence there is an asymmetrical shifting of the energy levels and the observed spectral line splitting, as illustrated in *Figure 2.3 (b)* and *(c)*, where the typical Zeeman and Stark effects for atomic sodium are shown side by side.

However, when the case of a gaseous molecule is concerned the situation is somewhat different. Thus most gas molecules are in a $^1\Sigma$ state and hence possess no resultant permanent magnetic moment, except for a very small contribution from the nuclei and molecular rotation. There are a few exceptions to this, such as oxygen and nitric oxide, but the Zeeman effect produced by even a large magnetic field on a typical gaseous molecule is usually very small. On the other hand, the electric dipole moment of many molecules is quite large and arises from the asymmetry of the distribution of the electron clouds around the molecule as a whole. It follows that the application of relatively small electric fields of the order of only 50 V/cm, can sometimes produce a large splitting of the absorption line. The detailed theory of this Stark effect and the way it can be used to assist in the analysis of spectra is considered further in the next section, but before this is done, one very practical application of this Stark effect should be noted.

38

In most forms of spectroscopy one of the major requirements is usually higher sensitivity in the spectroscope, so that smaller quantities of the specimen can be analysed, or lower pressures be used to produce higher resolution of the fine splittings. In nearly every piece of physical instrumentation the limit on sensitivity can usually be traced to one particular component. Thus it is usually found that this will produce more background noise than other components in the system, and hence the search for higher sensitivity becomes concentrated on improving the characteristics of this particular item of the equipment. In the case of microwave spectroscopes this item is the crystal detector, which converts the microwaves at the end of the waveguide run into a d.c. or audio frequency, which is used to display the actual absorption line. It transpires that these crystal detectors have a large amount of excess noise, when operating at d.c. or audio frequencies, and hence one of the ways in which the sensitivity of any microwave spectroscope can be improved is to try to produce some modulation of the observed spectra at a high frequency, so that this excess low-frequency noise of the detecting crystal can be overcome.

Early attempts to operate detecting crystals in such a way involved high-frequency modulation of the microwave source itself, but this can produce spurious signals from reflections down the waveguide run, and it became clear that modulation techniques, which would only produce a signal when absorption from the specimen itself was taking place, would be inherently better. The existence of a Stark effect allows just such a modulation technique to be employed. Thus a central metal electrode can be placed down the middle of the waveguide absorption cell and insulated from the surrounding waveguide walls. A high-intensity electric field can then be applied between the centre electrode and the waveguide wall, so as to produce a Stark effect in the absorbing gas, which will in turn produce the splitting of the observed spectral line. All that this technique has produced so far is the splitting in the spectra rather than any higher sensitivity. If, however, this electric field is switched on and off at a high frequency of, say, some 100 kHz, then the microwave absorption will also be amplitude-modulated at the same frequency. The output from the crystal detector can then be fed to a 100 kHz amplifier which follows it, and from which the actual absorption line can be extracted, as illustrated in *Figure 2.4*. Although this spectrometer is a little more complicated then the simplest ones previously discussed, it does nevertheless have a much greater sensitivity because the detecting crystal is now operated well up in the higher-frequency region and hence away from its low frequency excess noise.

It is also possible, by employing what is known as phase sensitive

39

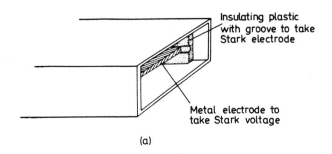

Insulating plastic with groove to take Stark electrode

Metal electrode to take Stark voltage

(a)

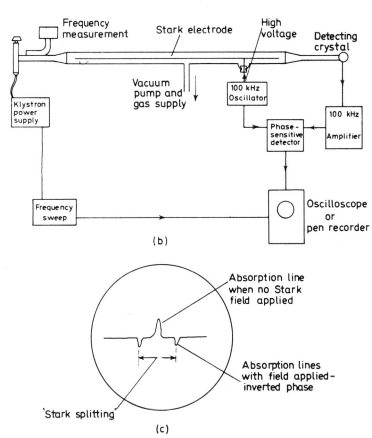

Frequency measurement

Stark electrode

High voltage

Detecting crystal

Vacuum pump and gas supply

100 kHz Oscillator

Klystron power supply

100 kHz Amplifier

Phase-sensitive detector

Frequency sweep

Oscilloscope or pen recorder

(b)

Absorption line when no Stark field applied

Absorption lines with field applied – inverted phase

'Stark splitting'

(c)

Figure 2.4. Spectrometer employing Stark modulation: (a) mounting of Stark electrode in waveguide; (b) block diagram of spectrometer system; (c) observed Stark splitting of spectral line

40

detection, to display the original absorption line on one side of the oscilloscope trace, and the Stark splitting components on the other, since the first of these will be obtained when the electric field is switched off, while the Stark components are only obtained when the electric field is switched on, as illustrated in *Figure 2.4*. It is clear that not only has higher sensitivity been achieved by this method, but also extra information such as the direct measurement of the molecular dipole moment is readily available. Other, more sophisticated, electronic means have also been used for increasing the sensitivity of gaseous spectroscopes such as superheterodyne detection, which employs a second microwave oscillator to produce a high-frequency beat note with the original signal, and hence overcomes the low-frequency noise of the detecting crystal. The rapid application of these techniques from the work of the microwave radar engineers is quite a fascinating study of the interaction between applied and pure research but there is no space for detailed consideration here.

2.3 Rotational Spectra and Chemical Structure

It has already been seen in the first section of this chapter that a diatomic or linear triatomic molecule will have a simple rotational spectrum, and to a first approximation the frequency of the observed lines is given by equation 2.4. Since the parameter B depends inversely on the moment of inertia of the molecule, it is clear that this can be deduced immediately from the observed spectra and hence, if the atomic weights of the individual atoms are known, which is normally the case, the separation between them, and hence the bond lengths in the molecule, can also be readily determined. The relation between these moments of inertia and the number of lines to be expected in the microwave region can be simply illustrated by taking the two cases of cyanogen bromide (BrCN) and hydrocyanic acid (HCN).

The BrCN molecule has a moment of inertia about its perpendicular axis of 204×10^{-47} kg m^2. Therefore, according to equation 2.4, frequencies of its rotational line spectra should be

$$\nu = 2[6.624 \times 10^{-34} / 8\pi^2 \times 204 \times 10^{-47}] . J = 0.8226 . 10^{10} . J \text{ Hz}$$

Hence it follows that the absorption line corresponding to its first rotational transitions will occur at 8226, 16 452 and 24 678 MHz respectively and there will be 36 absorption lines between 16 000 and 300 000 MHz (i.e. in the 2 cm to 1 mm wavelength region). In contrast to this the spectrum of HCN has only three lines in the same frequency range, since its moment of inertia is much lower, being only 18.9×10^{-47} kg m^2, and the frequencies of its first two rotational transitions are therefore 88 800 and 177 600 MHz.

As expected, it is the diatomic and linear molecules containing the heavy atoms which have most lines in the microwave region and therefore it is these which have been analysed in more detail by these techniques. In fact, if the moment of inertia of the molecule is less than 15×10^{-47} kg m^2 none of its rotational lines will come into the microwave region, and all will occur at wavelengths of less than 1 mm.

Equations 2.2 and 2.3 are only correct to a first order and have to be modified when more precise determinations are made. These modifications arise from the centrifugal distortion which occurs as the molecule starts to rotate faster and faster. Such centrifugal motion will tend to force the atoms further apart and thus alter the moment of inertia of the molecule as a whole. This fact had already been taken into account in the early infrared studies and a second-order correction for the energy levels can be made and equation 2.2 rewritten in the form

$$\frac{E}{h} = B.J \ (J + 1) - \frac{4B^3}{\omega^2} J \ (J + 1)^2 \qquad (2.6)$$

The second term now represents the correction for centrifugal distortion and it can be seen that it varies with the fourth power of J and also inversely as ω^2, where ω is the fundamental vibrational frequency of the molecule. It will be remembered that this vibrational frequency is, in its turn, governed by the elastic restoring force in the molecule and this will determine the amount of centrifugal distortion which takes place in the molecule under conditions of high rotation. It will be appreciated that precise measurements of the spectral line frequencies for higher values of J will thus enable information to be obtained on the elastic restoring forces as well as on the bond lengths themselves.

It should be mentioned, however, that there is also another effect which can sometimes modify the simple expression for the energy level and this is known as 'l type doubling'. This rises from the interaction of the rotational energy with the bending motion which can also tend to distort the normal configuration of the atoms and hence produce a change in the rotational energy levels. This interaction has the effect of producing a small doublet splitting of the observed lines and can usually be identified as such.

The rotational spectra to be expected from diatomic linear molecules are therefore relatively simple, consisting of a series of nearly equally spaced lines, the separation between each being approximately equal to $h/4\pi^2 I$. Hence I can be determined from the measured line frequencies and for a diatomic molecule the bond length can be calculated directly from I and the nuclear masses as explained above. The one determination of I is not sufficient to give internuclear distances in the case of triatomic linear molecules, however, and it is then necessary to replace some of the nuclei by different isotopes and thus obtain several different values of I, as the mass at different points along the molecule is changed. This method of isotopic substitution has been used to determine the bond lengths of a large number of linear polyatomic molecules in this way.

Some of the second-order effects, referred to above, can be observed very strikingly by employing spectroscopes which use the harmonics of the actual microwave radiation obtained from the oscillator. Thus any klystron which is feeding a waveguide run with its own fundamental frequency, can be used to produce harmonics of this frequency by passing its fundamental onto a crystal detector which then radiates the harmonics into a waveguide of smaller size. A large number of harmonics, ranging up to twenty times the original fundamental frequency or more, can then pass down the second waveguide through the absorbing gas, and the absorption lines can then be detected in the normal way at the end of a microwave run. It might appear that this technique has the disadvantage that the absorption lines corresponding to each different harmonic frequency would tend

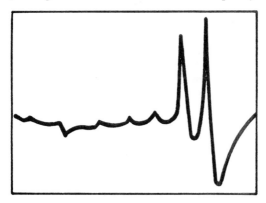

Figure 2.5. Second-order effects observed by use of harmonics. The 8th, 10th, 12th, 16th, 18th and 20th (from right to left) rotational lines of OCS obtained with the 4th to 10th klystron harmonic. Frequencies range from 97 to 243 kMHz

to interfere and overlap with one another. Indeed, if the absorption frequency is given precisely by equation 2.4 then all the absorption lines for the different harmonics would lie exactly on top of each other since they would all be separated by a frequency equal to the fundamental. However, the second-order terms of equation 2.6 will separate them and the net result is as shown in *Figure 2.5* which shows a whole series of rotational lines observed for OCS. It can be seen that the eighth, tenth, twelth, fourteenth, sixteenth, eighteenth and twentieth rotational lines are all obtained on the one sweep and arise from the frequencies of the fourth to tenth harmonic of the 1.25 cm klystron passing down the waveguide at the same time. The effect of the centrifugal distortion is shown up very dramatically in such a spectrum since it is only the centrifugal distortion term of $16B^3 J^3 /\omega^2$ which separates the different lines. Thus they no longer have a constant difference in frequency between them and the time base on the figure thus corresponds to differences of the absorption line frequency from the nearest value of $2 B.J$ and is not just a linear frequency scale. It is evident that this is a very direct method for determining the distortion constant and such measurements can be related directly to the restoring forces within the molecule and to their detailed structure.

The microwave spectra associated with molecules which are more complex than diatomic or linear are somewhat more complicated, as might be expected, but can nevertheless often be analysed in a very precise way and details of their structure also determined. Thus symmetric top molecules may be taken as the next more complicated type of structure and these molecules have equal moments of inertia about two of the three principal axes. The rotational motion of these molecules is such that, while the whole molecule rotates about its main axis of symmetry, this axis itself precesses about the direction of the total angular momentum vector. Because the electric dipole moment of the molecule lies along its main axis, only changes in the precessional energy will cause absorption of the microwave radiation. The energy of the resultant quantum states can then be shown to be of form

$$\frac{E}{h} = \frac{h}{8.\pi^2 I_b} \cdot \left[J(J+1) \right] + \frac{h}{8\pi^2} \left[(\frac{1}{I_a} - \frac{1}{I_b}) K^2 \right] \quad (2.7)$$

where I_a is the moment of inertia along the main symmetry axis and I_b is the moment of inertia about each of the other two perpendicular principal axes. J is the total angular momentum quantum number, as before, while K is a new quantum number which measures the component of angular momentum resolved along the main axis. It is clear from the above equation that equal positive and negative values of K will produce the same energy values, although these energy levels can be split by the application of external fields.

44

As in the case of the diatomic and linear molecules, the above equation will only apply precisely to a rigid molecule, whereas in practice centrifugal distortion of the rotating molecule, which increases with rotational energy, will add second-order terms to the expression of the energy. These can be added to equation 2.7 and the small additional terms will, in turn, produce small splittings on the observed spectra in the same way as these have been noticed in *Figure 2.5* for the linear molecule. The detailed expression for these will not be followed up in this book but it will be evident that very precise information on the molecular structure and internal molecular binding forces can be deduced from the second-order splittings which can be determined from the observed spectra in this way. An example of the chemical information which can be deduced from such spectra is afforded by some of the first studies which were made on the methyl halides. From such studies a systematic comparison of the different carbon–halogen bond lengths was possible and it was discovered that the CF bond length was very short in methyl fluoride. This can be linked with the large amount of double bond character which is associated with this particular bond and such information, together with very accurate determination of both bond angles and bond lengths, as well as detailed data concerning the higher vibrational states, have allowed very precise analysis of molecular structure and properties to be obtained from such microwave spectra.

In addition to the straightforward analysis of the rotational motion of the molecule as a whole, very fascinating studies of interaction between different rotating groups within one molecule can be undertaken. Thus if a molecule contains two or more groups, which can rotate against each other about a common molecular axis, a mode of torsional vibration becomes possible and this is usually termed 'hindered rotation'. The presence of this type of motion sometimes affects the microwave absorption spectrum to a considerable extent and its presence can be used to determine two very important molecular parameters, i.e., the height of the potential barrier to the internal rotation and the presence or absence of a component of the molecular dipole moment at right angles to the axis of the rotating groups.

The first spectrum actually observed in microwave spectroscopy from a molecule possessing hindered rotational levels was that from methyl alcohol and the definite allocation of some of these lines to such hindered rotation was suggested by their linear Stark effect which is discussed in the next section. Microwave gaseous spectroscopy, in fact, provides the most powerful method at present available for studying this phenomenon of internal rotation, and detailed analysis of the different spectra can be made to determine the heights of the different potential barriers within the molecule to these different types of internal motion. The details of such an analysis, which become fairly complex, are beyond the scope of this book but it will be appreciated that measurements on the internal force constants within the molecule can be extremely important when deducing its precise structure and properties.

2.4 The Stark Effect and its Applications

The general principle of the Stark effect has already been discussed briefly in Section 2.2 and it may be noted that what is called a 'first-order Stark effect' will occur when a molecule already possesses its own permanent electric dipole moment, along the same direction as that of its total angular momentum vector. Energy splittings which are equal to the product of the dipole moment and the applied electric field will then be obtained and these will produce typical frequency separations in the absorption line of some 100 MHz for fields of the order of 100 V/cm.

In a large number of cases, however, the molecule will not possess a component of its electric dipole moment along the direction of its total angular momentum vector and then a first-order Stark effect will not occur, because no interchange of rotational energy can take place via two vectors coupled at right angles. In these cases only a second-order Stark effect splitting will be produced and this arises from the same effect as described briefly for a free atom, i.e. from the fact that the applied external electric field can itself induce a dipole moment in the molecule. Such an induced dipole moment will have a component in the direction of J, and hence a small Stark splitting, will also be produced by the interaction of this induced dipole moment with the field itself. As in the case of the free atoms, however, the energy shift will be proportional to the square of the applied field and the quantitative value of the splitting will be very much smaller than that of the first-order effect. It will be evident that there is a very close parallelism here between first-and second-order Stark effects and paramagnetism and diamagnetism which are produced by the interaction of an applied magnetic field. As for the case of diamagnetism, it also follows that the second-order Stark effect is usually only observed when the first-order effect is absent.

If the simple cases of diatomic or linear molecules are considered it is evident that in these cases the electric dipole moment must lie along the internuclear axis, since all the distribution of charge occurs along this one direction. On the other hand, the angular momentum vector will always be perpendicular to this axis since the molecule will be rotating in a plane which has its normal perpendicular to the internuclear axis. It follows from this that no linear molecule can have a first-order Stark effect. It can nevertheless have a second-order Stark splitting as discussed above. Calculation of the actual energy splittings produced by the second-order effect shows that the energy changes can be written in the form

$$\Delta E = \frac{\mu^2}{2hB} \left[\frac{J(J+1) - 3M^2}{J(J+1)\,(2J-1)\,(2J+3)} \right] E^2 \qquad (2.8)$$

where μ is the permanent dipole moment and M is the 'magnetic' quantum number defining the component of angular moment resolved along the direction of the applied field, E.

The crucial points about this expression are that the splitting of the sub-levels is proportional on the one hand to the square of the applied field, and on the other to the moment of inertia of the molecule.

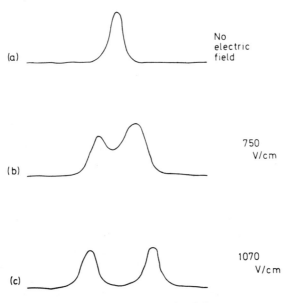

Figure 2.6. Stark splitting of OCS Spectrum

The actual magnitude of 'second-order' Stark splittings is very much smaller than the first-order effect, and a field of 100 V/cm will only

47

produce a splitting of about 0.2 MHz for small values of J and significantly less for the larger values. Thus quite high voltages have to be applied to the internal Stark electrodes if second-order Stark effects are to be studied in detail. One of the earlier Stark splittings observed can be seen in *Figure 2.6* which shows one of the rotational lines of the OCS molecule being split into two, by an electric field which rises to over 1000 V/cm

In contrast to the linear molecules the symmetric top molecules will possess a component of their electric dipole moment along the direction of the total angular momentum vector and its magnitude is given by

$$\mu K [J (J + 1)]^{-\frac{1}{2}}$$

Consequently unless $K = 0$, the first-order Stark splitting of the energy levels will occur and the magnitude of this splitting can be very readily calculated. Thus the component of the dipole moment along the direction of the electric field will be given by

$$\frac{\mu K}{[J (J + 1)]^{\frac{1}{2}}} \cdot \frac{M}{[J (J + 1)]^{\frac{1}{2}}}$$

and since the energy level splitting is equal to the product of this and the applied electric field, it will be given by

$$\Delta E = \frac{\mu.K.M}{J (J + 1)} \cdot \mathbf{E} \tag{2.9}$$

Hence the actual splitting observed in the spectral line for a transition obeying the normal selection rules of $\Delta K = 0$ and $\Delta J = \pm 1$ will be given by

$$h. \Delta \nu = \frac{2\mu.K.M.\mathbf{E}}{J (J + 1) (J + 2)} \tag{2.10}$$

It is evident that the moment of inertia of the molecule does not enter into this expression and the magnitude of the splitting will, in fact, only depend on the magnitude of the permanent electric dipole moment and the strength of the applied electric field. As mentioned earlier, this first-order splitting is very much larger than the second-order effect and will produce well-resolved lines even with small applied electric fields.

No new molecular parameters enter into the above expressions for the different types of Stark splitting, and it might therefore, be assumed that the study of such Stark splittings would not give any extra information about the molecule. They have, nevertheless, proved extremely useful in identifying different absorption lines and the symmetry of the molecules under study.

2.5 Nuclear Interactions and Hyperfine Structure

Some of the earliest studies on microwave gaseous spectroscopy had shown that, in addition to the splittings due to interactions with the molecular motion as a whole, much finer splittings could sometimes be observed on the spectra which were due to interactions with the individual nuclei within the molecule. These hyperfine splittings, as

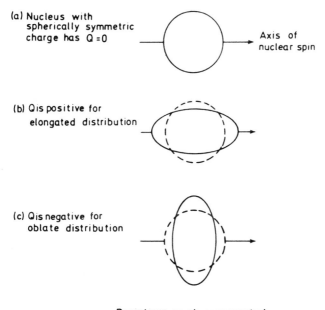

(a) Nucleus with spherically symmetric charge has $Q = 0$

Axis of nuclear spin

(b) Q is positive for elongated distribution

(c) Q is negative for oblate distribution

Deviations much exaggerated

Figure 2.7. Electric quadrupole moment. The axis is identified by the direction of the nuclear spin. The quadrupole moment, Q, is positive for a charge distribution which is elongated along the axis (b), and is negative for an oblate distribution (c)

they are called, arise from the interaction of the electric quadrupole moment of the nucleus and the asymmetric electric field which exists within the molecular structure. Thus the nuclei which are present in

49

the atoms of the molecule may possess their own spins, and associated with these there may be a magnetic dipole moment and an electric quadrupole moment. It is possible for both of these to interact with molecular rotational energy and therefore cause further energy shifts which vary with the orientation of the nuclear moment. However, the molecular magnetic field is usually very small and hence the splitting of the energy levels by this magnetic interaction is normally negligible. On the other hand, the gradient of the electric field at the nucleus is often quite large and the splitting of the energy levels by the coupling of this inhomogeneous field to the electric quadrupole moment of the nucleus is often sufficient to produce a well-resolved hyperfine structure.

The electric quadrupole moment of the nucleus is usually denoted by Q and is a measure of the deviation in the charge distribution from spherical symmetry. Thus a completely spherical nucleus would have a zero quadrupole moment, whereas one that has its charge in an ellipsoidal distribution will have a positive or negative quadrupole moment as the charge distribution is elongated or compressed along the direction of the axis of spin, indicated in *Figure 2.7*. The magnitude of the electrical quadrupole moment Q can be defined in terms of the equation

$$Q = \frac{1}{e} \int (3z^2 - r^2). \, dq$$

where dq is an element of charge in the nucleus having coordinates x, y and z and $r^2 = x^2 + y^2 + z^2$. It follows from such a definition that Q itself will have the dimensions of m^2 and is, in fact, of the order of magnitude of 10^{-20} m^2. The energy changes produced by the hyperfine interaction arises from the potential energy of this quadrupole moment as it orientates in the field produced by the outer molecular electrons. The detailed theoretical treatment of this interaction is fairly complex, but it can be shown that for a linear molecule the hyperfine interaction energy is given by an expression of the form

$$\Delta E = \left(\frac{\partial^2 V}{\partial z^2} \right)_{Av} . \; eQ. \; \frac{3/8 \; G(G+1) - 1/2 \; I(I+1) \, J(J+1)}{I(2I-1)(2J-1)(2J+3)} \quad (2.11)$$

where $G = F(F+1) - I(I+1) - J \cdot (J+1)$ and F is the quantum number of the total angular momentum inclusive of nuclear spin. The first term represents the gradient of the molecular electric field.

In a large number of molecules there is, of course, more than one nucleus possessing a quadrupole moment, and hence hyperfine interaction may well be observed from two nuclei interacting with the molecular electric field at the same time. One very good example of this, which was studied in detail in the early work on gaseous spectroscopy, is that of the ClCN molecule. This molecule is a good example of a two-nuclei interaction case in which one interaction is small compared with the other and can be regarded as a perturbation upon it. Thus, the spectrum under low resolution only shows three broadish lines, these being due to the quadrupole interaction with the chlorine

(a)

(b)

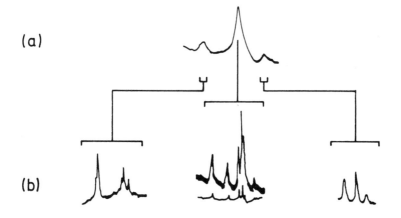

Figure 2.8. Hyperfine splitting in ClCN: (a) Spectrum under low resolution giving splitting due to chlorine quadrupole moment only; (b) Spectrum under high resolution showing extra splittings due to interactions with nitrogen quadrupole moment

nucleus and are seen at the top of *Figure 2.8.* At high resolution, however, each of these three lines will split into a collection of smaller lines and these extra components are now due to the small interaction with the quadrupole moment of the nitrogen nucleus and are shown in detail at the bottom of *Figure 2.8.* The predicted theoretical positions of the various lines can then be compared with those actually observed and a wealth of information is available from the analysis of such a detailed hyperfine pattern. Study of these spectra have not only allowed measurements to be made on the nuclear quadrupole moments themselves but also, once these are determined, it is possible to probe quite precisely into the electric fields existing within the various molecules.

51

2.6 The Advent of the Ammonia Maser

As mentioned already, one of the exciting features of developments in microwave and radio-frequency spectroscopy has been the rapid interchange between pure and applied science and the way in which these have helped each other with new ideas and new techniques. One very good example of this is the way in which the first masers, devices which have come to have enormous practical applications, arose from the work of the pure scientist in microwave spectroscopy. The term 'maser' is an abbreviation for microwave amplification by stimulated emission of radiation, in the same way that the word 'laser' stands for light amplification by stimulated emission of radiation. It is clear from these two names that the crucial principle involved in these new devices must be linked with the concept of stimulated emission of radiation, and a brief summary of the different ways in which radiation and matter can interact is really necessary before the operation of these new devices can be fully appreciated.

It has already been seen that the emission or absorption of radiation in any region of the spectrum corresponds to a transition between two energy levels. In the case of the visible region the two energy levels would be involved with the outer electronic structure,

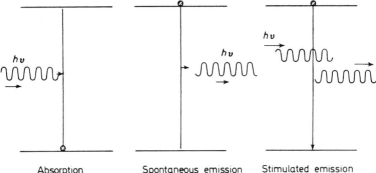

Absorption	Spontaneous emission	Stimulated emission
Incident quantum raises electron to higher level	Electron reverts to ground state - emitting quantum in process	Incident quantum stimulates electron already in higher level to emit second quantum -in phase with first
(a)	(b)	(c)

Figure 2.9. Interaction of radiation with matter: (a) absorption; (b) spontaneous emission; (c) stimulated emission

whereas in the infrared or microwave region they might be rotational energy levels of the molecule as a whole. In either case these two energy levels can be represented by the two horizontal lines in *Figure 2.9.* and the normal process of absorption is represented in *Figure 2.9. (a),* where an incoming quantum of energy of the right frequency is shown being absorbed by the atom, or molecule, to raise it to the higher energy level state as shown. After the atom or molecule has been excited to this higher level it can then spontaneously return to the lower level after a certain interval of time, and this type of emission of radiation in *Figure 2.9 (b)* is known as *spontaneous emission.* The crucial point about such spontaneous emission is that, although there may be an average time for the atom or molecule to exist in the excited state, there is no precise moment when it will return to the ground state and, moreover, the phase of the radiation which it emits is completely uncorrelated with the phase of any other radiation which is emitted from neighbouring atoms undergoing similar spontaneous emission.

There is, however, another type of emission which can take place instead. This is represented in *Figure 2.9 (c)* and is called *stimulated emission.* Such stimulated emission will occur if a second incoming quantum of radiation, of the same frequency, arrives whilst the atom or molecule is still in the excited state. It is then impossible for this new quantum to be absorbed, since the atom or molecule is already in the excited state, but this incoming quantum can stimulate the atom or molecule to emit its energy in the form of a second quantum of radiation. This will be emitted before the corresponding spontaneous emission, and this quantum of stimulated radiation will have its wavetrain in phase with that of the incoming quantum, so that the two now leave the system in step with one another, as illustrated diagrammatically in *Figure 2.9 (c)*. This is really the crucial point about stimulated emission because it provides a mechanism whereby one incoming quantum of radiation can produce two outgoing quanta which are in phase with one another or, in other words, are *coherent.*

The concept of coherent radiation has proved extremely important in recent years and can probably be best illustrated by considering the simple example of normal radio waves, as they are broadcast from a transmitter. The electric and magnetic fields, which form the electromagnetic radiation as it is propagated outwards from such a transmitting aerial, are themselves produced by the flow of electrons up and down the aerial. Thus, one can imagine pulses of current flowing up and down the aerial itself and, in so doing, giving rise to the changing electric and magnetic fields which produce the electromagnetic wave. For a continuous and uninterrupted sinusoidal wavetrain to be produced it is, of course, necessary for the current flow in the aerial to be of a regular periodic nature. If the individual electron pulses were moving up and down the aerial in a random way then the electric and magnetic fields which they produce would have no correlation between them and no coherent wave patterns would be obtained, It is relatively easy, however, to ensure that the current pulses in the aerial, and thus the wavetrains which they produce, do remain correlated and coherent by driving these from a master oscillator circuit. This can also be used to feed other aerials and produce a whole array from which the wave-trains are all in step and coherent. This situation is illustrated in *Figure 2.10 (a)* where four small dipole aerials are shown, each fed from the same oscillator and, therefore, each producing wavetrains which are in step with one another and producing a totally coherent pattern. If, on the other hand, these four separate aerials were all fed by separate oscillating circuits which had no connection with one another, then the wavetrains would be completely incoherent and unrelated as shown in *Figure 2.10 (c)* and the addition of them would produce no recognisable pattern.

This case of radio-wave production from oscillating circuits can now be compared with the radiation patterns that would be expected from emitting atoms or molecules. In the case of ordinary spontaneous emission from such atoms, or molecules, a completely uncorrelated set of waves will be obtained, as illustrated in *Figure 2.10 (d)* where four atoms are shown producing spontaneous radiation of an incoherent type. If, however, an incoming incident quantum can be used to stimulate these four atoms so that they all radiate at the same time

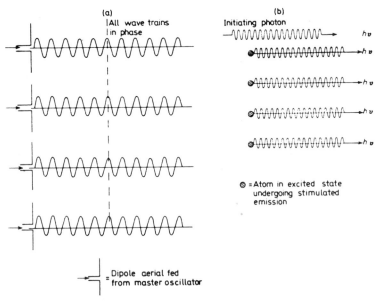

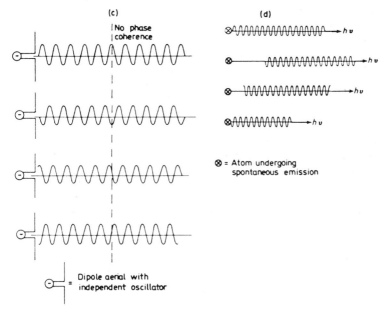

Figure 2.10. Coherent and incoherent radiation: (a) Coherent wave trains from four dipole aerials; (b) Coherent wave trains from atoms undergoing stimulated emission (laser action); (c) Incoherent wave trains from unrelated dipole aerials; (d) Incoherent wave trains from atoms undergoing spontaneous emission (normal visible radiation)

and phase, then the situation will be as shown in *Figure 2.10 (b)* and it can be seen that this is analogous to the four aerials driven by a master oscillator and a totally coherent wavefront can be produced.

One of the great advantages of such coherent radiation is that it enables weak signals to be amplified. Thus, if the weak signal is represented by the original incident quantum of radiation, it is clear from *Figure 2.10 (b)* that this can quadruple itself by interacting with the four excited atoms or molecules. These can, of course, go on in turn and interact with other excited atoms so that a large number of emerging quanta are produced, all in phase, and all coherent with one another. This is the basic principle of amplification by stimulated emission and is the underlying principle of the 'maser' and the 'laser'.

It follows that such a process of amplification can take place if there is a large number of atoms, or molecules, in the excited state. This is not the normal situation in any system, however, since in any collection of atoms or molecules in thermal equilibrium, there will always be more in the lower state than in the higher state. As a result,

55

any incoming quantum of radiation will have a higher probability of being absorbed than of producing stimulated emission and hence no successive effect of stimulated emission, as envisaged in *Figure 2.10 (b)*, could occur. Therefore, the normal energy level population must somehow be inverted if such stimulated emission is to take place effectively, and the crucial problem in designing any 'maser' or 'laser' system is to develop some method whereby the inversion of energy level populations can be produced.

It is at this point that there is a direct link with the work on microwave gaseous spectroscopy since the first method of achieving such inversion of energy level population and employing it for practical purposes was in the ammonia 'maser'. The population inversion was obtained by an actual spatial separation of the excited and ground state molecules, those in the ground state being dispersed away from the entrance to a cavity resonator, whereas those in an excited state were concentrated into a small beam and fired into the cavity which was tuned to the appropriate frequency. The possibility of carrying out such a physical separation of the molecules arises from the Stark effect of the ammonia molecule, and the way in which the two energy

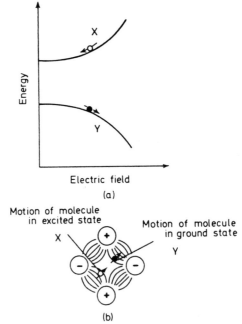

Figure 2.11. Stark splitting of NH$_3$ molecules and their spatial separation:
(a) Divergence of energy levels in an applied electric field;
(b) Field distribution between four wires of the separator;

56

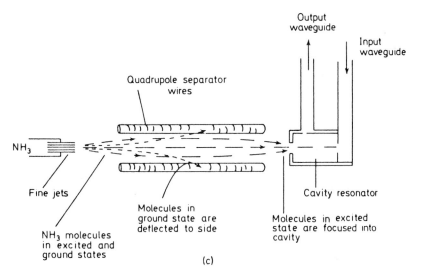

(c) Focusing of excited molecules into microwave cavity

levels of the molecule change when an external electric field is applied is shown in *Figure 2.11 (a)*. The energy value of the excited molecules (X) rises as the electric field strength increases, whereas that of the lower level (Y) falls. Thus, if an inhomogeneous electric field is produced over the region in which the ammonia molecules are moving, those in the excited state will tend to move towards the region of lower electric field strength, whereas those in the lower state will tend to move towards regions of higher electric field strength. If an electric field distribution such as that produced between four parallel wires, shown in cross section in *Figure 2.11 (b)*, is employed, then the low values of electric field will be in the centre of this region while the high values are to be found between the wires themselves. Hence, if a beam of ammonia molecules is fired down between these four wires, in a direction perpendicular to the plane of the diagram, the molecules in the excited state will be focused towards the axis while those in the ground state will be dispersed away from the centre towards the position between the wires. As a result, the higher energy and lower energy molecules are actually separated one from the other, and at the end of this separating device the excited molecules can be directed through a small hole into a cavity resonator, as indicated in *Figure 2.11 (c)*.

By using this kind of apparatus it is possible to produce a high concentration of excited ammonia molecules in the cavity resonator

57

and hence, produce conditions under which an incoming quantum of radiation will produce a large amount of stimulated emission rather than become absorbed. These are, of course, just the conditions required for amplification to take place and hence, one application of such an ammonia maser is to produce amplification of microwaves at this particular frequency. It can also be operated as an oscillator since any amplifier can be converted into an oscillator by simply feeding back some of the output to the input. The device will then oscillate at the resonance frequency for which such feedback occurs and in this particular case, it will be the resonance frequency of the ammonia molecule. Thus, if the energy population inversion is sufficiently large, the stimulated emission which will be initiated by a noise phonon of the correct frequency, will be sufficient to overcome the losses inherent in the cavity; and microwave oscillations can thus be obtained from the cavity, their frequency being determined by the inversion frequency of the ammonia molecule.

Such an oscillator provides an extremely precise frequency standard, since the frequency of the oscillations are determined entirely by the atomic and molecular constants of the ammonia molecule itself. The ammonia maser, operating as an oscillator in this fashion, was the first practical atomic clock based on such molecular oscillations, and with other similar atomic standards it now represents a more precise frequency and time standard than that offered by astronomical time. This is a good example of one of the very practical applications which fed back from the work of pure research to give immediate benefit to the microwave engineers, who had made microwave spectroscopy possible in the first place.

However, as well as these practical applications, the advent of the ammonia maser, and the maser principle in general, has also had a very practical effect on spectroscopy itself since it allows very much higher resolution spectrometers to be constructed than would otherwise be the case.

2.7 High Resolution Spectroscopy

It will be appreciated that when ammonia molecules are undergoing stimulated emission in the cavity of the ammonia maser as shown in *Figure 2.11 (c)*, they are in fact producing emission spectra instead of the normal absorption spectra. Moreover, the width of these emission lines can be made very much narrower than the normal absorption lines because the normal Doppler broadening associated with the random motion of the molecules in all different directions is very much reduced. Thus, the molecules can be observed along a line which is

58

perpendicular to their line of flight so that their motion in the direction of the line of observation is very small and can produce a Doppler broadening of only about 1 kHz. The very first studies of the ammonia lines obtained from such ammonia masers indicated this possibility since the 3.3 inversion line was found to have its main quadrupole hyperfine transitions further split into four finer components, these being produced by the interaction of the hydrogen magnetic moments with the molecular magnetic fields; and it has already been noticed that these magnetic interactions are very weak.

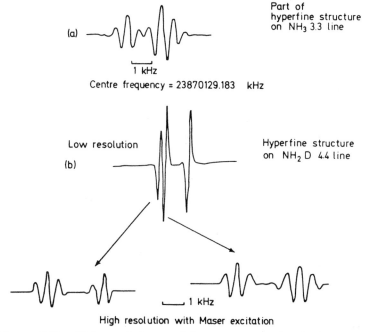

Figure 2.12. High resolution emission spectra produced by maser excitation: (a) part of hyperfine structure on NH_3 3.3 line; (b) hyperfine structure on NH_2D 4.4 line

The kind of resolution that can now be obtained in such emission spectra are illustrated in *Figure 2.12. Figure 2.12 (a)* shows the splitting of the ammonia hyperfine lines referred to in the last paragraph, while spectra from NH_2D are shown in *Figure 2.12 (b)*. The very fine splittings that can be observed under such conditions allow very detailed checking between the theoretical model of the molecular structure and the observed experimental results, and thus act as a very fine probe for molecular structure determination.

59

Chapter Three

Magnetic Resonance at Microwave Frequencies

3.1 The Principles of Electron Spin Resonance

The basic principle of electron spin resonance has already been considered in the first chapter and illustrated diagrammatically in *Figure 1.7.* It was seen there, that the unpaired electrons are separated into two distinct groups by the application of an external magnetic field and these two groups have a difference in energy equal to $g\beta B$. The electrons in the lower level are then excited up to the higher level by absorption of microwave radiation of the correct resonance frequency and hence an absorption line is obtained. The essential elements of any electron resonance spectrometer are, therefore, as already shown in *Figure 1.8,* and include a source of the microwave frequency in the form of radar valve, and a cavity resonator which concentrates this radiation on to the specimen which is also placed at the centre of a strong magnetic field.

After the basic principles and techniques had been discussed briefly in the first chapter it was pointed out that, in a large number of cases, hyperfine splitting might be observed in the spectrum and the single absorption line would then be replaced by a hyperfine pattern. The existence and analysis of these hyperfine patterns has provided some of the most important and interesting features of electron resonance spectroscopy and the way in which it arises can be explained briefly as follows.

Such hyperfine splitting arises from the interaction of the unpaired electron with the nucleus around which it is moving. In this case the interaction is of a magnetic nature and the field due to the electron couples with the magnetic moment of the nucleus itself; thus, only nuclei which possess nuclear spins and magnetic moments will produce any such hyperfine patterns. So far in our consideration of the energy levels of the unpaired electron, we have only considered the interaction of the electron with the externally applied magnetic field. There may, however, be magnetic fields which arise within the atom or molecule containing the unpaired electron and these will need to be considered as well. The most obvious source for such an internal magnetic field will be the nucleus of the atom around which the electron is moving. To illustrate this specifically we could take the case of a copper atom which has one unpaired electron in its 3d-orbital and this will be moving around the copper nucleus, which itself has spin and magnetic moment. The magnitude of this nuclear magnetic moment is very much smaller than that of the electron spin, because the mass of the nucleus is very much larger than the mass of the electron and magnetic moments are inversely related to the mass of the particle which possesses them. However, the distance between the nucleus and the electron is also very small and it is, therefore, possible that the field produced by the nuclear magnetic moment at the site of its own electron might be quite significant. In fact, a simple classical calculation shows that the value of such a field for a copper nucleus is about 10^{-2} tesla (i.e. 100 gauss). This additional field will produce a very noticeable effect on the observed spectrum and it is clear that such hyperfine structure may be a very important feature of the spectrum.

The nuclear spin, I, of the copper nucleus is 3/2 and hence it can orientate in four different directions, when placed in an external magnetic field. Thus, it can align itself either with, or against, the direction of the applied field to give resolved quantum numbers of $M_I = +3/2$ or $M_I = -3/2$. However, it is also possible for the copper nucleus to take up two intermediate positions as shown in *Figure 3.1 (a)* and these intermediate positions will have resolved quantum numbers of $M_I = \pm 1/2$. The essential quantum condition in all of these cases is that the appropriate quantum numbers should differ by 1 in their successive values. These four orientations of the nuclear spin and magnetic moment will in turn produce four different incremental values of magnetic field at the site of the electron, and these are represented diagrammatically in *Figure 3.1 (a)* by the corresponding letters. These four incremental magnetic fields will, therefore, change the simple energy level diagram which has been previously illustrated in *Figure 1.7* and split each of the original electronic levels into four as shown in *Figure 3.1 (b)*. An incoming quantum of microwave energy will now find four different values of the external field which will satisfy the resonance condition, as shown in the figure, and it can be seen that in some cases the internal fields from the nucleus assist the external field, whereas in others they oppose it. The single absorption line is thus split into four components, these four components corresponding to the four orientations of the copper nucleus. This is just a particular case of a more general relation which states that a nuclear spin of value, I, will produce $(2I + 1)$ equally intense hyperfine lines, this being equal to the number of ways in which it can orientate in an applied field.

The spectrum which is actually obtained from a simple inorganic copper salt is shown in *Figure 3.1 (c)* and the four component lines can be clearly seen. It is evident that they are double with the two outer lines clearly splitting, but with the inner lines still unresolved. This doubling arises from the fact that copper contains two abundant isotopes ^{63}Cu and ^{65}Cu, both of these have a nuclear spin of 3/2 but they have slightly different magnetic moments and hence produce slightly different splittings. It may be noticed that there are also some other very weak lines between the two outer sets of hyperfine splittings and these arise from the non-spherical shape of the nucleus and the fact that it has an electric quadrupole moment. This particular feature will not be considered in any more detail here, but such a spectrum does illustrate very clearly the high resolution

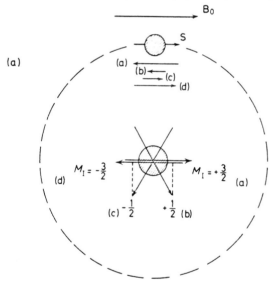

Figure 3.1. Hyperfine interaction with a copper nucleus: (a) The (2I + 1) different orientations of a copper nucleus with their associated incremental fields;

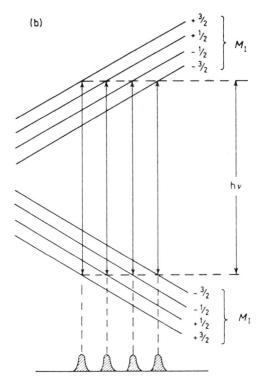

(b)

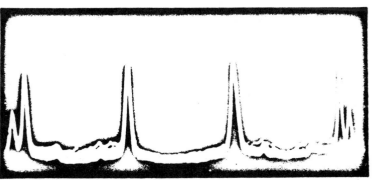

(c)

(b) Resultant splitting of the two electronic levels and the four field values for resonance; (c) Observed hyperfine pattern. The doubling arises from the Cu^{63} and Cu^{65} isotopes which are both present.

63

that can be obtained by working in this region of the spectrum, and the wealth of information that can also be deduced from second-order splittings in such a spectrum.

The above paragraphs describe in detail what happens when the unpaired electron experiences the magnetic field due to one nucleus. In a large number of cases, however, and especially in chemical applications such as for free radicals or other organic compounds, the unpaired electron will be moving in a highly delocalised orbital and its wave function may well embrace several different nuclei instead of just being linked to one. One therefore has to consider what type of hyperfine structure will be produced if the unpaired electron interacts with several nuclei at once. Two particular cases might be considered in detail initially, the first is where the electron interacts equally with several identical nuclei, as for example with several protons, and the other extreme case will be when there are two groups of nuclei and the interaction with one group is very much stronger than with the other.

As a specific example of the first type of interaction one might take the hyperfine splittings which are produced from a CH group and compare them with a CH_2 group, or with a CH_3 group. As the normal isotope of carbon does not possess a nuclear magnetic moment, no splitting will be obtained from it, and consequently only the interaction with the protons need be considered. In the first case

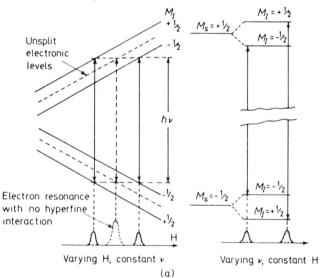

Figure 3.2. Hyperfine patterns from equally-coupled protons:
(a) Doublet produced from one proton;

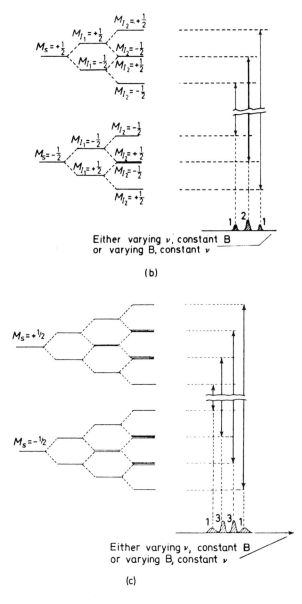

(b) Triplet produced from two equally-coupled protons; (c) Quartet produced from three equally-coupled protons

65

the unpaired electron will experience the additional field produced by the single proton of the CH group. This proton has a spin of $I = \frac{1}{2}$ and will either be aligned with or against the applied field, therefore, it can only take up two possible orientations with the resultant quantum numbers $M_I = \pm \frac{1}{2}$. Consequently a doublet structure will be observed as a hyperfine pattern and this is just one particular case of the general $(2I + 1)$ rule which was noticed previously. This particular energy level and resultant hyperfine pattern can be represented as in *Figure 3.2 (a)* where it is seen that the original electronic levels are split into two, corresponding to two different orientations of the proton, and a doublet splitting is actually produced on the observed electron resonance spectrum.

If, however, the electron is interacting with the two protons of a CH_2 group then the interaction with one of these will produce a doublet splitting of the original electronic levels, just as described above for the CH group, but the further interaction with the second proton must now also be considered. This will produce a further splitting in each of the levels which have already been formed, and if the coupling to the two protons is equal the second splitting will be of equal magnitude to the first. This is shown in *Figure 3.2 (b)*. It is evident that the net result of these two splittings is to bring the central pair of resultant energy levels together so that there are now only three distinct energy levels for each electronic state, but the centre one of these is formed from two which have coalesced. Its population will, therefore, be twice as great as either of the others, and the net result is that there will be a hyperfine splitting into three components with intensity ratio of $1 : 2 : 1$. This splitting is also shown diagrammatically in *Figure 3.2 (b)*.

It is evident that the interaction with the three protons of the CH_3 group can be analysed in the same way. Thus, the interaction with the two protons will produce a triplet splitting in exactly the same way as already discussed for the CH_2 group, and then each of these will now split again by the interaction with the third proton. In a CH_3 group the magnitude of the interaction with all the protons will be the same and hence this last splitting will be the same magnitude as for the previous two and the net result of this can be seen in the energy level diagram of *Figure 3.2 (c)*. It is seen that each of the electronic energy levels are now split into four, but the centre two are composed of three individual components which have all come together. As a result of this the hyperfine pattern observed will be of four lines but with intensity ratio of $1 : 3 : 3 : 1$, as is indicated at the bottom of the figure.

It will be evident that this type of reasoning can be extended indefinitely and a simple mathematical formula can be built up for the hyperfine pattern to be expected for n equally coupled protons. It can be shown that $(n + 1)$ hyperfine lines will be obtained and intensity ratios of the components of this pattern will form a binomial series. The observed hyperfine structure from an unpaired electron moving in a molecular orbital embracing several nuclei is thus different, both quantitatively and qualitatively, from that observed from a single nucleus, and although the particular case of protons has been considered in some detail the same argument can, of course, be extended to any nuclei possessing magnetic moments.

It was mentioned earlier that the other extreme case which might be considered is when there are two interactions present which are of very different order of magnitude. If the simple case of an unpaired electron interacting with two nuclei is considered, and the interaction with the first nuclear spin, I_1, is much greater than that with the second nuclear spin I_2, then the interaction with the first will produce $(2I_1 + 1)$ hyperfine lines which are fairly widely separated, while the interaction with the second nucleus will then produce a much smaller splitting of each of these original lines into a further $(2I_2 + 1)$ components. If the second interaction is very small compared with the first it is probably that no overlapping of these different sets of superhyperfine lines will occur. Many examples will be found of such cases where each original hyperfine line is split into a number of additional components by further interactions with ligand nuclei. The analysis of such superhyperfine structures has enabled a great deal of information to be obtained on the detailed mechanism of exchange interaction in the solid state as well as serving as a fine probe of molecular structure.

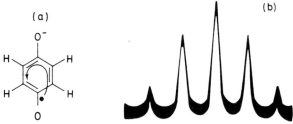

Figure 3.3. Observed hyperfine patterns from protons in semi-quinone: (a) Structure of benzosemiquinone radical; (b) Observed five-line hyperfine pattern

In the same way that the hyperfine structure actually observed from a single nucleus, such as copper, has been illustrated in *Figure 3.1*,

67

so the observed hyperfine pattern from four equally-coupled protons can be illustrated as in *Figure 3.3*. This spectrum is from the benzosemiquinone free radical which consists of a benzene ring in which two of the six protons have been replaced by oxygen atoms and an unpaired electron has been released from one of these to move around the whole ring structure as illustrated in *Figure 3.3 (a)*. It is evident from considerations of the pure symmetry of such a molecule that the unpaired electron must interact equally with all four protons and hence, if the argument in the above paragraphs is true, a five-line hyperfine structure is to be expected with a binomial distribution in the intensity of the lines. The actual hyperfine pattern observed is shown in *Figure 3.3 (b)* and it can be seen that this agrees very well with that predicted.

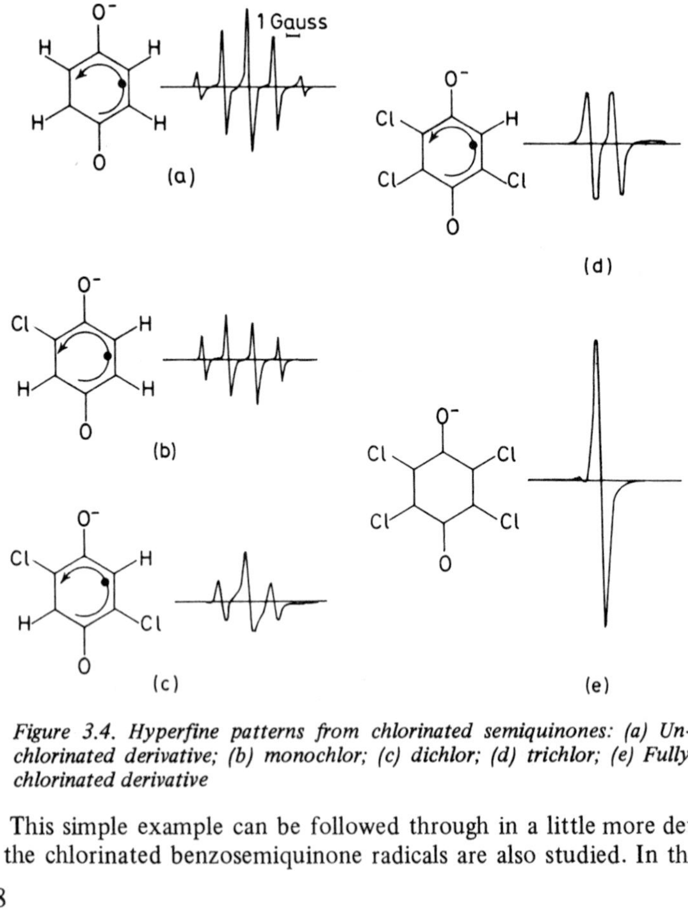

Figure 3.4. Hyperfine patterns from chlorinated semiquinones: (a) Un-chlorinated derivative; (b) monochlor; (c) dichlor; (d) trichlor; (e) Fully chlorinated derivative

This simple example can be followed through in a little more detail if the chlorinated benzosemiquinone radicals are also studied. In these each proton can be replaced in turn by a chlorine atom and as this is done the number of hyperfine components is successively reduced, until only a single line is observed, indicating that all four protons have been replaced by chlorine atoms. The actual spectra observed from such a series of chlorinated benzosemiquinones is shown in *Figure 3.4*, (*a*) being that of the unchlorinated derivative with its four protons, while (*b*), (*c*), (*d*) and (*e*) trace the series through from the monochloro to the dichloro and trichloro derivatives. It is evident that in each case the *n* remaining protons produced (*n* + 1) hyperfine lines and the predicted intensity rations of 1 : 3 : 3 : 1 and 1 : 2 : 1 can be clearly seen.

It may be noted that the spectra illustrated in *Figure 3.4* are presented as first derivative recordings and not as the absorption lines of *Figure 3.1* and *3.3*. In all modern work on electron resonance, and indeed in a large number of other types of spectroscopy, it is usual to find the spectra presented as such first derivatives since this is the normal way in which high sensitivity electronic display equipment produces the final output. Thus, each absorption line is represented as a swing from one side, across the axis to the far side and back, and the centre of the absorption line is the actual crossover point on the axis itself. The particular reasons why absorption lines should be presented in this way, and the methods whereby they can be analysed are related to the development and design of the spectrometer systems themselves which are now considered in a little more detail.

3.2 Electron Resonance Spectrometers

The simplest type of electron resonance spectrometer has already been shown in *Figure 1.8*, but it will be appreciated that this will also suffer from the low frequency excess noise properties of the detecting crystals discussed in Section 2.2 since the detected currents are flowing at very low audio frequencies. Hence, if higher sensitivity is to be obtained in such a spectrometer this low frequency crystal noise must be overcome in the same way that it was overcome in the gaseous spectrometers by using Stark modulation at higher frequencies. Exactly the same principle can be used in the case of magnetic resonance by modulating the magnetic field applied to the specimen at high frequency and then detecting this modulation on the microwave power at the output crystal. This high frequency field modulation is now used in most commercial electron resonance spectrometers and can produce very sensitive and robust systems. The frequency used for such magnetic field modulation is usually about 100 kHz, since an additional form of broadening on the spectral lines, known as modulation broadening, will be produced if too high a frequency of modulation is used.

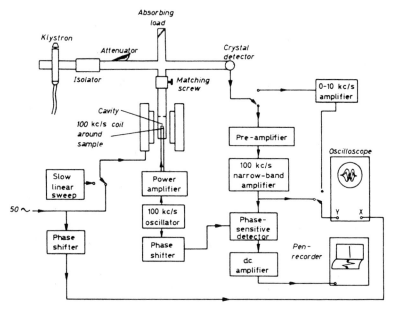

Figure 3.5. Block diagram of high frequency modulation ESR spectrometer

A typical modern electron resonance spectrometer using such high frequency field modulation is shown as a block diagram in *Figure 3.5.* It is seen here that the 100 kHz oscillator feeds a small coil, which is placed around the specimen inside the cavity resonator, and this coil produces the high frequency magnetic field modulation in parallel with the main d.c. magnetic field. The microwave absorption is, therefore, also modulated at this frequency and this modulation is detected by the crystal at the end of the microwave run and fed to a narrow-band 100 kHz amplifier from which the final absorption spectrum can be obtained, either on an oscilloscope screen, or on a pen-recorder as shown. The noise content of the detecting system can be reduced by employing a technique known as phase-sensitive detection, in which a reference signal from the 100 kHz oscillator is also fed to the amplifier so that the signal is only amplified when it is in phase with the modulating field. The actual depth of modulation produced by such a coil is not large and usually it is smaller than the width of the spectral line. As the main d.c. magnetic field passes through the resonance value so this small high frequency modulation rides over the profile of the absorption, sampling its slope at any point as it does so. It is for this reason that the actual result plotted out by the pen-recorder is in the form of the first derivative of the absorption line rather than the simple absorption curve itself.

Various other techniques have been developed over recent years to increase the sensitivity of such spectrometers and one of these, which has very general application, employs a 'computer of average transients', or C.A.T. for short. This is able to store successive runs through an absorption line and then, by integrating these together, it is possible to lift a very weak signal out of the background noise, and hence improve the sensitivity of the spectrometer by many orders of magnitude. Such techniques are now finding increasing application, not only to improve the sensitivity of ordinary spectro-meters, but also to allow study of rapidly changing signals which may be produced by photolysis or rapid chemical change. Further details of these methods are discussed in Section 3.7. They are another very good example of the rapid interchange between the work of a pure scientist and of the applied electronic engineer.

One other basic technique which has now found a place in electron resonance spectrometry and especially in connection with its application to biochemical studies should be mentioned, however. This is known as the continuous flow method and the general idea is illustrated in *Figure 3.6*. It can be applied to any system in which two solutions are to interact and these are stored in the reservoirs R_1 and R_2. The outlets from these reservoirs meet in the mixing chambers M, and the outlet from this leads through a reaction tube K. This tube can be made of quartz and is suitable for insertion into a standard rectangular cavity of the spectrometer. The distance x, along K before the tube enters the microwave cavity, can be altered and in this way the time between the initiation of the reaction and the passing of the mixture into the cavity can also be adjusted. If unlimited amounts of the two reacting solutions are available, they can be allowed to flow into the mixing chamber at suitable rates and the electron resonance spectrum of the resultant products can then be followed, as a function of time, by gradually moving the quartz reaction tube further and further out of the cavity (i.e. increasing x). One of the main uses of this kind of equipment is to study various enzyme reactions in biochemistry when the appearance and disappear-ance of both free radical intermediates and changing valency states of transition group atoms can be followed at the same time.

One great disadvantage of the above technique, however, is that it

71

does use rather large amounts of the reacting solutions and this may be very difficult to obtain for some of the more specialised reactions to be studied. This particular disadvantage can be overcome by what is known as the 'sudden freezing technique' which is really an extension, or adaptation, of the flow method described above. This is illustrated diagrammatically in *Figure 3.6 (b)* and it is seen that

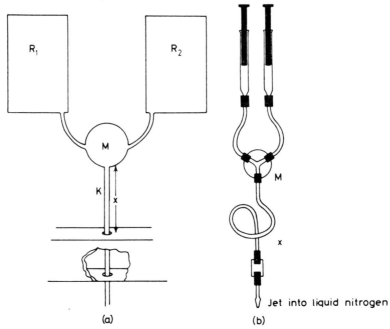

Figure. 3.6. Continuous flow systems for transient studies: (a) General principle of continuous flow systems. The reacting solutions are mixed in the chamber, M, and the time between this mixing and entering into the ESR cavity can be varied by altering the distance, x, or the rate of flow down the tube; (b) Sudden freezing technique. The reacting solutions are now instantaneously deep frozen instead of passing straight to the ESR cavity

the reacting solutions are still driven from their reservoirs into a mixing chamber and then on through a reacting tube. However, this tube does not pass through the electron resonance cavity but instead ends in a fine jet through which the reacting solutions are injected into a very cold liquid so that they immediately become deeply frozen and the reaction stops. The most satisfactory way to achieve this appears to be the direction of the fine jet of reacting solutions into hexane cooled to liquid-nitrogen temperatures. It can be shown that under these kinds of condition a diluted aqueous solution, originally at room temperature, will only take about 10 ms to become deeply frozen with all its reactions stopped. The deeply frozen material will then fall to fill the narrow tail of the containing vessel and when this has become packed with the frozen material it can be inserted into the electron resonance spectrometer, the tube always being held at liquid-nitrogen temperature. The concentration of the intermediate species in the reaction can thus be studied at leisure and the method also has the advantage that much smaller quantities of reacting material are required. Moreover, sensitivity is also increased from the lower temperatures that are employed. Examples of how both of these techniques have very significantly helped in the application of electron resonance to biochemical studies are given later in Section 3.7.

3.3 Parameters of E.S.R. Spectra

There are five basic parameters which can be used to characterise an electron resonance spectrum. There are: (i) its integrated intensity, (ii) the width of the absorption lines, (iii) the g-value, or resonance field position, (iv) any electronic splitting which is present, (v) any hyperfine splitting which is present. Not all of these features will necessarily be present in every spectrum, for example the electronic splitting will only occur when there are two or more unpaired electrons associated with the atom or molecule under study. Similarly, hyperfine structure will only be present if the unpaired electron is interacting with a nuclear magnetic moment as explained previously. The first three parameters can be used to characterise any and every electron resonance spectrum, however, and a brief consideration of the information which can be deduced from them is necessary before the application of electron resonance to any specific group of compounds is discussed.

3.3.1 Integrated Intensity

The integrated intensity of the absorption line is, of course, a direct measure of the microwave power absorbed by the specimen and this, in turn, is directly proportional to the difference in population between the two levels concerned. At the fixed values of microwave frequency and operating temperature, this difference in population will be a constant proportion of the total number of unpaired electrons in the specimen. Hence it follows that the magnitude of the microwave absorption, and hence of the integrated intensity of the spectral lines, will be directly proportional to the total number of unpaired

73

electrons present in the sample. Thus, provided what is known as 'power saturation' is not taking place, the number of unpaired electrons present can be directly monitored from the integrated intensity of the absorption lines. This kind of information can often be of very direct interest when following transient species in a chemical reaction, or the growth and decay of free radicals, or radiation damage, and other similar applications.

3.3.2 Spectral Line Width

A broad shallow absorption line may have the same integrated intensity as a sharp narrow line, but the width at half height of a shallow line will be considerably larger than that of the narrow line. It thus follows that the width of the absorption line is an additional parameter from which further information can be deduced. It is clear that this linewidth will be determined directly by the energy spread of the two levels giving rise to the transition and hence the linewidth will be a measurement of, and give information on, the interactions which the unpaired electron is experiencing and which cause its energy to be spread in this way.

Such interactions can be grouped under two headings. The first is termed spin–lattice interaction and the second, spin–spin interaction. The term spin–lattice interaction embraces the various ways in which the unpaired electron can interact with its general surroundings, such as the crystal lattice in a solid, or the general molecular structure of an organic molecule. This interaction provides a mechanism whereby the extra energy taken up by the unpaired spins is returned to the surroundings as a whole. It is clear that such an interaction must be present if the resonance absorption is to continue, since otherwise electrons originally in a lower energy level would be raised to the upper level continuously until the two levels were equally populated when the absorption would cease. The spin–lattice interaction therefore provides the mechanism whereby saturation effects, which are discussed later, can be avoided and the general thermal equilibrium distribution of electrons between the two levels maintained. If, however, the interaction is very strong the excited electrons will remain in the upper level for only a very short period of time, Δt, and there will then be a spread in the energy associated with the upper level given by the uncertainty principle relation

$$\Delta E . \Delta t = h/2\pi \tag{3.1}$$

as previously discussed in Section 1.2.

It follows from this that if the time spent in the higher level is very short because of the very strong interaction with the lattice, there will then be a considerable spread of energy of the levels and thus broad absorption lines will be produced. For example if the lifetime is reduced to an average of 10^{-10} seconds the equivalent frequency spread will be about 10^9 Hz and thus, in terms of a magnetic field spread on a normal electron resonance absorption line, this would correspond to about 300 gauss or 0.03 tesla.

Such broadening of the absorption lines by spin–lattice interaction can nearly always be reduced by cooling the specimen to low temperatures, since the mechanism for such interaction involves a coupling to the thermal vibrations of the lattice or molecule. It is for this reason that the electron resonance spectra of some transition group atoms, such as Co^{2+} or Fe^{2+} can often only be observed at low temperatures, when resonance lines which are broadened beyond detection at room temperature can be reduced in width to such an extent that clearly resolved lines are obtained.

The other type of interaction known as spin–spin interaction covers all the mechanisms whereby the electron spins can exchange energy amongst themselves rather than give it back to the lattice or molecular system as a whole. It follows that such interactions cannot assist in the general establishment of thermal equilibrium, but they will nevertheless produce a broadening of the resonance line, both from the direct magnetic interaction of the spins on each other, and from the smaller lifetime of the spin states which they can also induce. Such a direct magnetic interaction between the spins can be considered as equivalent to the classical interaction between two small bar magnets, and for a single pair of spins will vary in angle as $(3 \cos^2 \theta - 1)$, which is characteristic of the field produced by any magnetic dipole. Such a variation can often be clearly observed in the experimental results and it will be seen a little later that exactly the same form of angular variation also occurs for the electronic splitting and one major form of hyperfine interaction. It should also be noted, however, that in solutions the very rapid tumbling motion of the molecules can average out this angular variation to zero. Thus, any given unpaired electron attached to a specific atom may be rapidly moved through all the possible positive and negative values of such a $(3 \cos^2 \theta - 1)$ variation, in a time which is short compared with the inverse of the absorbing frequency.

It is obvious from these brief qualitative comments that the study of the width of the absorption lines, and its variation with temperature, may often give very direct information on the interactions which are taking place within the specimen under study. It should also be noted that it is not only the width of the line but also its shape that can give

75

clues on the different interactions which are present. Thus, the normal spin—spin dipolar type of interaction will produce a line shape of Gaussian form, while the line shape produced by most spin—lattice type of interactions, and by the exchange or averaging effects discussed above, is of a Lorentzian form which is considerably narrower in the centre than a Gaussian shape, but spreads out with more intensity in the wings.

3.3.3 g-values

The other main parameter which can be associated with a single absorption line is its g-value, as defined from its magnetic field resonance position and the equation

$$h\nu = g.\beta.B \qquad (3.2)$$

It follows from this that if the microwave frequency is held constant, which is normally the situation in any spectrometer system, then the only two parameters which can vary are the value of the externally applied magnetic field, B, and the magnitude of the g-value. Hence a measurement of the resonance field value will determine the g-value to be associated with the particular unpaired electron, and this will, in turn, give information on the chemical binding or structure of the atom or molecule in which the electron is moving.

An electron with no orbital angular momentum will have a g-value exactly equal to that of the free electron spin, i.e. 2.0023. A large number of organic free radicals do in fact have g-values very close to this magnitude, since their electrons are moving in highly delocalised molecular orbitals and have very little interaction with individual atomic orbitals. If, on the other hand, the unpaired electron is closely associated with a single atom it may possess considerable orbital angular momentum and this will shift the g-value well away from that of the free spin. The basic reason for the shift in g-value is that the magnetic moment associated with spin angular momentum is twice that associated with the equivalent orbital angular momentum. This fact produces what is known as the Landé splitting factor which is introduced in ordinary atomic spectroscopy to explain the Zeeman effect.

Such a free-atom g-value can be calculated from a straightforward vector model in which magnetic moments of $\beta \, [L \, (L + 1)]^{1/2}$ and of $2\beta \, [S \, (S + 1)]^{1/2}$ are drawn along the directions of the angular momentum vectors which couple to give the overall total angular momentum vector $h/2\pi \, [J \, (J + 1)]^{1/2}$. The net atomic magnetic moment can then be determined trigonometrically by evaluating the projection of the total magnetic moment along the direction of $h/2\pi \, [J \, (J + 1)]^{1/2}$.

76

The value obtained in this way is given by

$$\text{Effective atomic magnetic moment} = g\beta[J(J+1)]^{1/2}$$

where $g = 1 + \dfrac{J(J+1) + S(S+1) - L(L+1)}{2J(J+1)}$ (3.3)

The theory of such g-values is not quite so simple when the atoms under study are contained within a solid crystalline lattice or large molecule, however. There are then strong internal electric fields acting on the atom and the unpaired electron cannot be considered as just bound to one single individual atom, as is assumed in the theory of atomic spectroscopy. These internal electric fields act on the orbital states of the atoms in question and can very significantly change their energy levels and, in so doing, also alter the g-values which will be observed for the electron resonance spectra. The precise details of the theory behind such g-values in the transition group atoms is beyond the scope of an introductory volume of this type, but it will be evident that the association of different g-values with different transition group atoms will serve as a very powerful identifying characteristic and this can often be assisted by the angular variation which is observed in the spectrum, if single crystals are studied. Examples of the way in which such information can be deduced from measurements of g-values are given in the later sections of this chapter.

3.3.4 Electronic Splitting

In all considerations so far it has been assumed that only one unpaired electron is associated with any atom or molecule under study and this is indeed often the case. It is, however, possible for two or more unpaired electrons to be present, and the formation of a triplet state in an organic molecule by photolysis provides one good example. In such a case as this, one of the electrons in the normal covalent pair bond can be excited to a higher level so that the two spins no longer pair but have their spins parallel. There will then be two unpaired electrons associated with this molecule and their spins will couple to give a resultant spin quantum number of $S = 1$ for the molecule as a whole.

The same situation can also exist in transition group atoms where lower levels, such as the $3d$, are filling up beneath a higher level which already holds a paired group of electrons, such as the $4s$ in the first transition group. In such cases the number of unpaired electrons associated with an individual atom may vary from one (for the case of Ti^{3+} or Cu^{2+}) to five (for the cases of Mn^{2+} and Fe^{3+}) in the first transition group. The simple energy level diagram of *Figure 1.7*

77

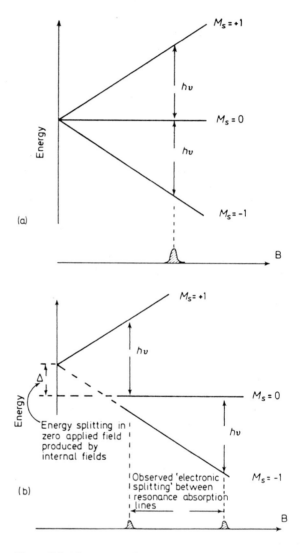

Figure 3.7. Electronic splitting for S=1: (a) If no internal electric field the two absorption lines coincide; (b) If internal electric field is present it produces an initial splitting and the two absorption lines are now separated

78

will, therefore, no longer apply in these cases and, since such cases do form a fairly high fraction of the transition group atoms studied, a brief consideration of this electron splitting is necessary.

It will be simplest to start with the consideration of just two unpaired electrons in the triplet state of an organic molecule, or in the atomic orbitals of a transition group atom. The combined total spin quantum number of $S = 1$ can now orientate as a single unit in the applied magnetic field and there will now be three different orientations which are possible corresponding to the magnetic quantum numbers $M_s = + 1, 0$ or -1.

If the particular molecule or atom were in free space, and not subject to any internal electric fields, all these three orientations of the total spin would have equal energies in a zero applied magnetic field. When an external field was applied their energies would separate as shown in *Figure 3.7 (a)*. Thus, the level corresponding to $M_s = 0$, and hence zero magnetic moment along the field axis, will be unchanged in energy, whereas the two corresponding to the $M_s = \pm 1$ will move up and down in energy as shown. If the microwave frequency is now fed to such a system it is clear that the resonance absorption will occur between the +1 and 0 level at exactly the same value of magnetic field as for the transition between the 0 and -1 level. Hence, although two transitions are being excited by the incoming microwave frequency, only one absorption line will be observed because the two transitions exactly overlap. In practice, however, the unpaired electrons will not be in such a free space condition but contained within either a complex organic molecule or a crystalline lattice. In either case they will be experiencing strong internal electric fields which will act on the electrons and produce a Stark effect, in the same way as discussed earlier for the gaseous molecules when the field was applied externally. Hence, even in the absence of any applied magnetic field there will now be a splitting in energy between the $M_s = 0$ and $M_s = \pm 1$ levels, as illustrated in *Figure 3.7 (b)*, this splitting being due to, and a measure of, the internal electric field. If the microwave frequency is now applied it is evident that absorption will take place at two different values of the applied magnetic field, as shown in *Figure 3.7 (b)*. As a result two absorption lines, instead of one, are observed and the splitting between them is itself a direct measure of the electronic splitting between the levels of the atom or molecule in question. Measurements of such electronic splittings can often give very useful information on the internal interactions present within the molecule or crystal lattice as a whole and, as will be seen in Section 3.6 on three-level masers, such electronic splittings can also often have considerable practical application.

79

3.3.5 Hyperfine Splittings

The way in which hyperfine splitting arises from the magnetic interaction of the unpaired electron with the nuclear magnetic moment has already been discussed in some detail in earlier sections of this chapter. It was seen there that interaction with one single nucleus will produce equally intense hyperfine lines, whereas coupling to more than one nucleus will produce a 'christmas tree' or binomial distribution in the intensities. No consideration has been given yet, however, to the actual mechanism whereby such interaction takes place, and it is rather important to have some understanding of this if the kind of information that can be obtained from such hyperfine splittings is to be clearly understood.

In earlier qualitative considerations a model of two interacting bar magnets, or dipoles, was used. It has already been seen that the spin–spin interaction responsible for the broadening is of the same basic nature, and it therefore follows that both interactions will have the classical dipolar angular variation of $(3 \cos^2 \theta - 1)$. The spin–spin dipolar interaction between neighbouring unpaired electrons will be much greater than that between an electron and nuclear moment, and it is often necessary to dilute any concentrated paramagnetic specimens with diamagnetic material to move the unpaired electrons further apart if hyperfine structures are to be observed. Otherwise the weaker interaction between the electron spin and the nucleus will be obscured by the much stronger interaction between neighbouring unpaired electrons. Such hyperfine splitting, with this angular variation of $(3 \cos^2 \theta - 1)$ is in fact often observed when such diluted single crystals are studied and a large amount of useful information can be deduced about the symmetry and chemical nature of the atoms surrounding the paramagnetic centre from such studies of angular variation.

However, this simple dipole–dipole interaction cannot be the only interaction present which produces hyperfine splitting, since otherwise no hyperfine lines would ever be observed in solutions. Thus, the rapid motion of molecules in a solvent will average out the angular interaction for the hyperfine structure just as it does for the spin–spin broadening. In a large number of cases quite large hyperfine splittings are observed from solution and, since there are no reference angles in such cases, it is clear that these splittings must arise from an interaction which is independent of angle and therefore cannot be of the normal dipole–dipole type. This new interaction, which is known as the Fermi or 'contact' interaction, has no analogy in classical physics and arises from the fact that an unpaired electron moving in an s-orbital has a finite probability of actually being at the site of the nucleus of its atom. Moreover, s-orbitals in a wave-mechanical picture are entirely spherically symmetric and therefore have no angular variation associated with them. Consequently, if an s-orbital contains an unpaired electron, it is possible for a completely isotropic splitting to arise from the interaction between its magnetic moment and that of the nucleus and it is this 'contact' type of interaction which produces all the splittings that are observed from specimens in solution. In the case of solid or crystalline specimens both types of hyperfine interaction will be present and will need to be differentiated by careful analysis of the angular variation. The detailed correlation of the measured hyperfine splittings and those predicted by various theories has been one of the most fruitful testing grounds of theoretical physics and chemistry. Detailed analysis of these splittings has provided very precise probes both of the crystal lattices in solid state physics, and of complex molecular structure in organic chemistry. Specific examples of these are to be found in the sections which follow.

3.4 Studies of Transition Group Atoms

Although there are a wide variety of different types of compound that can be studied by electron resonance, there are, nevertheless, two main groups. These are the free radicals of physical and organic chemistry on the one hand, and the transition group complexes of inorganic chemistry and solid state physics on the other. The analysis of the e.s.r. spectra of the free radicals is generally much simpler than that of the transition group atoms since there is very little coupling to the orbital angular momentum. Moreover, the free radicals are usually studied in solution, and hence only the isotropic hyperfine splittings need to be considered. The spectra obtained from transition group atoms, on the other hand, can be quite complex and all the five parameters listed above often need to be considered and their angular variations carefully analysed. Although it often takes longer to analyse such spectra, it follows as a corollary that more information is eventually available on the atoms and systems concerned, and a brief description of such transition group complexes is therefore now summarised.

Various theories have been proposed to explain the energy levels of an atom inside a solid and one of the most successful of these was the crystal field theory which regards the transition group atom as located in an electric field, the symmetry and magnitude of which are determined by the dipolar charges of the surrounding atoms or groups. The energy levels, which are normally degenerate for a free atom, are then split by this internal field and the way in which this splitting takes place can be deduced from general group theory. Once the splittings between the various orbital levels of the atom have thus been obtained, the expressions for the g-values which will be observed experimentally can be deduced in terms of the ratio of this splitting to the spin–orbit coupling parameter of the free atom. As a general comment, it may be said that the larger the ratio of a spin–orbit coupling to the splitting between the orbital levels, the more the observed g-value will depart from the free spin value of 2.0023. The details of such theoretical calculations are somewhat too complex for a book of this nature, but it will be clear that the measurement of g-values does give direct information on the energy states and chemical binding of the atom in the solid.

However, such a simple crystal field theory on its own cannot explain all the results observed in the transition group complexes and another approach via the molecular orbital method was then developed. In the case of transition group complexes, this method is applied to the complex formed by the transition group atom and its immediate ligand atoms rather than to the whole molecule, as in organic chemistry. The molecular orbitals are then determined by combining orbitals on the central paramagnetic atom with those on its ligand atoms, following the general rule that the symmetry of the orbitals on the atoms must match. Thus, for instance, for the first transition group atoms only certain combinations of the ligand orbitals will have the same symmetry as the d-orbitals on the central metal atom and it is only these which can be combined to give molecular orbitals for the complex itself. The great advantage of this approach is that it gives a direct explanation of the superhyperfine structure which can be observed from the surrounding ligand atoms.

In fact, neither the crystal field theory nor the molecular orbital theory by themselves can fully account for all the experimental results obtained by the electron resonance studies and a dual approach was therefore developed, known as the 'ligand field theory'. This concentrates on the energy level splittings produced for the complex as a whole and considers the energy level splittings as arising from a combination of effects, including the electrostatic effect of the internal field and those due to the π and σ bonding, which can be taken as the direct contribution of the ligand molecular orbitals. As an example of this approach, the interaction between the atomic orbitals on a central paramagnetic atom and those on the surrounding ligand atoms, can be considered. The unpaired electron of such a central paramagnetic atom will be moving in a d-orbital and the five different shapes of the d-orbitals available to such an unpaired electron are shown in *Figure 3.8*. It will be seen that all of these, with the exception of that along the z-axis, are basically of a similar configuration, in the shape of two dumb-bells, but orientated differently in space. The particular shape of the d_{z^2} orbital is due to the fact that the z-axis has been chosen as the axis of quantisation, and it will be clear that if all these five orbitals are summed together a totally spherically symmetrical charge distribution will be obtained for the complete electron shell.

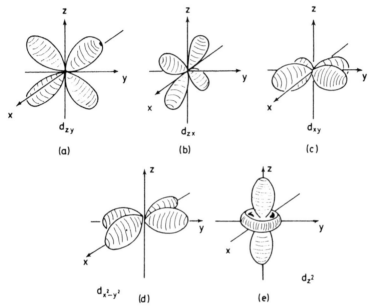

Figure 3.8. Shapes of 3d-orbitals of a free atom: The orientations of the five independent 3d-orbitals are shown with cases (a), (b) and (c) intersecting the three perpendicular axes, while (d) and (e) point along them

In the absence of any perturbing electric or magnetic fields, these orbitals all have the same energy, but energy level splittings will be produced if ligand atoms with their own electron clouds are now brought close to the central metal atom. The simplest case is where six ligand atoms surround the central paramagnetic atom at equal distances along the positive and negative x, y, and z axes, as shown in *Figure 3.9*. It is clear that the five different d-orbitals can now be grouped into two. In the first group are those which have their orbitals and therefore electron concentration pointing towards some of the surrounding ligand atoms (such as $d_{x^2-y^2}$ and d_{z^2} − orbitals), while in the second group are those which have their orbitals pointing in directions between those of the ligand atoms (such as d_{xy}-, d_{yz}-, and d_{zx}- orbitals). There is considerable electron repulsion between those in the first group and the ligand atoms and hence these are raised in energy, and two groups of energy levels now result with the $d_{x^2-y^2}$ and d_{z^2} levels noticeably higher than the other three.

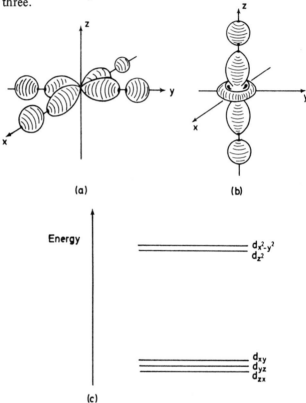

Figure 3.9. Central atom surrounded by octahedron of ligand atoms: (a) and (b) show how repulsive interaction takes place between the ligand atoms and the $d_{x^2-y^2}$ and d_{z^2} orbitals. This produces a shift of these orbitals to a higher energy level, as shown in (c)

In this derivation, however, it is assumed that the distribution of the electron density on each of the six surrounding ligand atoms is spherically symmetrical (i.e. the electrons are in s-orbitals). In

84

practice, it is much more likely that some of the ligand atoms will have p-orbitals associated with them especially if they are taking part in a π-bond system. These asymmetric orbits will now further affect the orbitals on the central paramagnetic atom and produce a further splitting of its energy levels. In *Figure 3.10* the different interactions between the p-orbitals on the four ligand atoms in the xy-plane are shown, and it is clear from this diagram that these will interact strongly with the d_{xz}- and d_{yz}-orbitals, raising them in energy, but not with the d_{xy}, and hence a further splitting of the levels is produced.

A detailed consideration of the p- or π-orbital on the atoms at the fifth or sixth co-ordination point, as shown in *Figure 3.10(c)* and *(d)*, shows that this lower group of levels can be further split. If one of the atoms at the sixth co-ordination point is taking part in a π-bond system, its p-orbital will be at right angles to the plane of the π-bond system, as indicated, and then this p-orbital will approach close to the yz-orbital if the orientation is shown as in *Figure 3.10 (d)*. Additional repulsion will therefore take place and the d_{yz}-orbital will be moved to a higher energy than the d_{zx}. If, on the other hand, the plane of the π-bond system is at right angles to that shown, then it would be the d_{zx}-orbital that would be moved higher in energy, rather than the d_{yz}. The relative spacing of such orbitals can be evaluated fairly directly from the experimental g-value variations and hence this information will immediately indicate the actual orientation of the π-bond system around the paramagnetic atom. A very good example of how this has been used in practice is the case of the haemoglobin derivatives, where such structural information on the symmetry surrounding the iron atom, obtained in this way from electron resonance measurements, helped in the actual elucidation of the structure of the molecule itself.

This molecular orbital approach also enables the interaction of the unpaired electron with the ligand atoms to be more directly visualised. Thus, if orbitals of the ligand atoms with the correct symmetry are envisaged as mixing with those on the central atom, the electron can be considered as moving out from the paramagnetic atom to share in the ligand orbitals and thus interact with their nuclear and magnetic moments as well. This delocalisation or sharing of the unpaired electron, which can also be interpreted as a significant amount of covalent bonding occurring in the complex, would, in general, produce four observable changes on the electron resonance spectrum i.e.:

(i) a reduction in the orbital contribution to the g-value and hence a reduction of its shift away from the free spin magnitude;

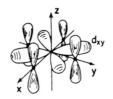

(a)

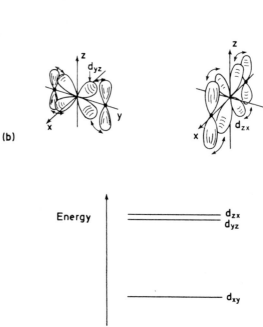

(b)

Energy splitting as a result of (a) and (b)

(c)

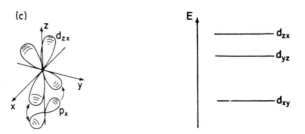

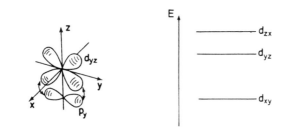

(d)

Figure 3.10. Interaction with p-orbitals of ligand atoms: (a) Interaction with d_{xy}-orbital is minimal; (b) Interaction of d_{yz} and d_{zx}-orbitals with p-orbitals of the four ligand atoms in xy plane is considerable. Hence these are raised in energy compared with d_{xy}; (c) and (d) Interaction of d_{zx} and d_{yz}-orbitals with p-orbital on sixth co-ordination point. The orientation of this orbital will determine which of the d_{zx} or d_{yz}-orbitals is further raised in energy

(ii) a reduction in the magnitude of the hyperfine splitting observed from the interaction with the central paramagnetic atom. This arises simply because the unpaired electron is not spending so much of its time interacting with the central atom;

(iii) an increase in the spin—lattice relaxation time, since this is also dependent strongly on the orbital interaction with the central atom;

(iv) the appearance of superhyperfine splitting from the magnetic moments of the ligand nuclei.

Various examples of such superhyperfine patterns are shown in *Figure 3.11*. The first two are from crystals of iridium chloride (a) at an angle when the interaction is only with the iridium, (b) with the chlorine h.f.s. as well. The very detailed measurements of the strength of these interactions which are now available from the hyperfine splittings allow the molecular orbitals to be calculated in great detail, and these superhyperfine splittings serve as an additional fine probe of such molecular and solid state interactions. They have been used with great success not only in the study of transition group complexes themselves, but also for defects in inorganic crystals, where such hyperfine patterns arise from the trapped electron interacting with the ligand atoms around it.

3.5 Saturation and Relaxation Effects

So far in these considerations only brief attention has been paid to the width of the spectral lines, and to the information that can be deduced from it. It has been seen that this gives details on the

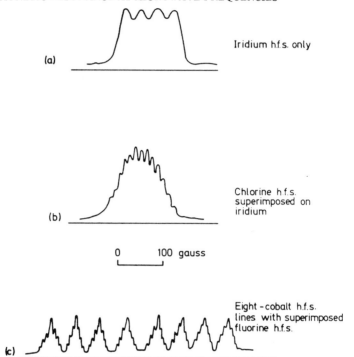

(a)   Iridium h.f.s. only

(b)   Chlorine h.f.s. superimposed on iridium

0   100 gauss

(c)   Eight-cobalt h.f.s. lines with superimposed fluorine h.f.s.

Figure 3.11. Super hyperfine patterns (a) and (b) from iridium chloride (c) from cobalt fluoride

interactions present within the solid state, but study of the different mechanisms of broadening has also been found to be of immense practical value. Thus, the existence of what is termed 'saturation broadening' can be linked directly with the invention and operation of the three-lever maser which is another of the new devices which the pure scientist has handed back to the applied engineer. In order to appreciate how the three-level maser works it is necessary first to consider in a little more detail saturation effects that can occur in electron resonance transitions.

In any spectroscopic transition two competing factors must be considered. On the one hand, the number of atoms in the excited state is increased by the presence of the incoming microwave radiation, but, on the other hand, this increase in population is reduced by the spin—lattice interaction, which takes energy from the excited state to share with the general vibrations of the lattice, and in so doing re-establishes conditions of thermal equilibrium. What is known as power saturation occurs when the spin–lattice interaction is not strong enough to return the excited atoms to ground state at a rate sufficiently fast to maintain the normal population distribution. If this happens the transition probabilities between the two levels will begin to change and reduction in intensity and broadening of the observed absorption line will result.

This state of saturation can be treated quantitatively by considering two energy levels, A and B, where A is the excited level and has a population of N_1 atoms, while B has a population of N_2. In thermal equilibrium and in the absence of any incoming radiation, the ratio between the energy level populations will be given by the normal Maxwell–Boltzmann distribution, i.e.

$$N_1/N_2 = \exp\left(-h\nu/kT_L\right) \tag{3.4}$$

where T_L is the temperature of the lattice. When the microwave resonance radiation is applied, however, transitions will be induced from the ground state to the excited state, via the normal process of absorption, and stimulated emission will also take place in the reverse direction. The net effect of both of these transitions can be represented by the term $(dn/dt)_{r.f.}$ and the total rate of change of the population can, therefore, be written as

$$\left(\frac{dn}{dt}\right)_{total} = \left(\frac{dn}{dt}\right)_{r.f} - \left(\frac{dn}{dt}\right)_{s.l.} \tag{3.5}$$

where $n = N_2 - N_1$ and $(dn/dt)_{s.l.}$ represents the action of the spin–lattice interaction in returning the excited atoms to the ground state. This can be written in terms of the deviation of the population distribution from normal, and the spin–lattice relaxation time, T_1, in the form.

$$\left(\frac{dn}{dt}\right)_{s.l.} = \left(\frac{n_0 - n}{T_1}\right) \tag{3.6}$$

where n_0 is the value of n at thermal equilibrium as given by equation 3.1.

The value of $(dn/dt)_{r.f.}$ can be obtained directly from classical radiation theory and is given as

$$\left(\frac{dn}{dt}\right)_{r.f.} = \tfrac{1}{4} \cdot \pi \cdot \gamma^2 \cdot (B_1)^2 \cdot g\left(\omega - \omega_0\right) \cdot n \tag{3.7}$$

where γ is the gyromagnetic ratio of the electron, B_1 is the strength of

89

the microwave magnetic field, and g $(\omega-\omega_0)$ is the shape function of the absorption line having been normalised so that its integral over all frequencies is equal to unity. Its value, at resonance, can be replaced by the magnitude of the spin–spin relaxation time which, in turn, determines the width of the line. Thus, it can be shown that the maximum value of g $(\omega-\omega_o)$ is equal to T_2/π.

When equilibrium has been re-established in the presence of the microwave radiation the total $(dn/dt)_{total}$ must be zero and hence

$$\tfrac{1}{4}. \gamma^2 . (B_1)^2 . T_2 . n = \frac{n_0 - n}{T_1} \tag{3.8}$$

Therefore

$$n/n_o = \frac{1}{1 + \tfrac{1}{4} \cdot \gamma^2 . (B_1)^2 . T_1 . T_2} = Z \tag{3.9}$$

This ratio of the population difference between the two levels, in the presence of the microwave power, to that under thermal equilibrium conditions is often referred to as the 'saturation factor', Z, the smaller the value of Z the higher the saturation that is taking place. It follows from the parameters in equation 3.9 that the absorption will be reduced by large values of the spin–lattice relaxation time, T_1, and by increase in the level of the microwave power, B_1. This decrease in the expected absorption obviously occurs first in the centre of the lines, where the greatest power is absorbed, and only affects the wings of the lines as the value of the microwave radiation rises still further. Thus the effect of saturation is not only to reduce the expected power absorbed but also to alter the line shape, flattening it in the centre before it does so in the wings, and thus increasing the apparent width. The onset of such saturation effects can therefore be readily observed from the spectra since the absorption lines will appear to broaden as the microwave power is increased.

The analysis of saturation, which is taking place within an electron resonance spectrum, can be of help in identifying the nature of un-resolved structure in the line. Thus, if the broadening is due to homogeneous effects, such as the spin–spin interaction between electrons on neighbouring atoms or motional and exchange narrowing, the whole line will act as a unity and broaden in the way described above. If, however, the broadening of the line arises from unresolved hyperfine structure, each individual hyperfine line will undergo the saturation broadening just discussed. In this case the envelope of all the lines will not change shape, but the expected intensity of absorption will fall equally across it. Hence, detailed study of the way in which the shapes of the lines change, as the integrated intensity is reduced under saturation conditions, enables the presence of unresolved hyperfine structure to be detected within a single broadened line.

However, the most significant application of saturation studies is in the more applied field where it forms an essential element of the design and operation of the three-level maser systems. These three-level masers again form a very good example of the way in which the pure scientist has, almost accidentally, handed a powerful new device to the electronic engineers in partial repayment of the debt which was owed to them for providing the tools to study this part of the spectrum.

3.6 Three-level Maser Systems

It has already been seen, when discussing the advent of the ammonia maser in Section 2.6, that the possibility of using atomic systems to

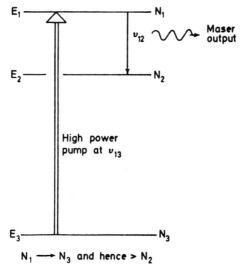

$N_1 \longrightarrow N_3$ and hence $> N_2$

Figure 3.12. Principle of three-level maser: High energy pumping between the ground state and higher level E_1 increases population of E_1 above that of E_2 and hence population inversion occurs, with the possibility of stimulated emission and maser action

amplify incoming microwave signals depends crucially on an inversion of the normal energy level population, so that incoming quanta can

91

produce stimulated emission rather than absorption. Methods of achieving this by actual separation of the excited and ground state atoms, as in the case of the ammonia maser, must involve very complex and sophisticated equipment. Inversion of the population distribution between two levels can be achieved within a solid, however, if a system of energy levels containing three or more components is utilised, and saturation is produced between two of these.

The principle of this three-level maser operation is shown in *Figure 3.12* and under normal thermal equilibrium conditions the population distributions will be such that $N_1 < N_2$ and $N_2 < N_3$. However, if a large amount of microwave power is now fed in at the frequency ν_{13}, saturation effects can take place between the ground state level E_3 and the higher level E_1, as indicated by the broad arrow, and under such conditions the population of these two levels may become nearly equal, so that N_1 is only slightly less than N_3, and significantly greater than N_2. In this way the population inversion between the upper two levels E_1 and E_2 can be achieved, so that an incoming microwave quantum of frequency ν_{12} will produce stimulated emission, rather than absorption, when it interacts with the atoms in these levels. It follows that one now has the conditions necessary for microwave amplification at this frequency, and, what is more, these conditions can be continuously maintained as long as a high-power pumping frequency ν_{13} is fed into the system, so that atoms are fed to the top excited level as fast as the stimulated emission removes them. It is clear that there has to be a careful choice of conditions for such three-level masers to operate. In the first place paramagnetic compounds which have energy level splittings of the right order of magnitude must be chosen, and in the second place, the conditions of operation must be such that saturation is relatively easily obtained between the ground state and a higher level. One of the great advantages of using paramagnetic salts, however, is that the energy level spacings can be changed by altering the value of the applied magnetic field, as already discussed when electronic splittings were described in sub-section 3.3.4. In this way they provide a ready-made atomic amplifier for use in the microwave region. Their great advantage, compared with other amplifiers, is that the only noise in the system is due to the fundamental thermodynamic noise and there are no large extra components, such as those associated with hot electron beams or point contacts, in valve or normal solid state devices. Thus, the three-level maser offers the microwave engineer an amplifier with ideal noise characteristics and hence much higher sensitivity than others available. They were the amplifiers which made the first transatlantic television relays a possibility.

From the experimental point of view, the three basic conditions that must be met in the operation of such a three-level maser device are: (i) a transition group compound with a suitable energy level system; (ii) the correct conditions for saturation to be effective must then be provided, and this normally requires cooling the specimen to a low temperature so that the relaxation time is lengthened, and saturation can take place at reasonable input microwave power levels. It should be noted that the low temperature has the added advantage that it will reduce the inherent noise of the amplifying system. (iii) it is then necessary to devise a microwave structure which can be resonant at both the pumping frequency and the frequency of amplification, but this does not produce any fundamental difficulties in cavity or co-axial resonator design.

Although the laser, as such, does not come within the terms of the microwave and radio-frequency region, it might be relevant to point out here that the principle of the laser is almost identical to that of the maser, and followed very rapidly after it. In the laser various means are used to throw atoms up to an excited state in exactly the same way as indicated in *Figure 3.12,* but with the difference that the energy splittings concerned are very much bigger; and optical transitions rather than microwave frequencies are involved. Once an inverted. population has been obtained between two of the higher levels, however, stimulated emission can take place between them and a laser can thus be operated, either as an amplifier, or, more frequently, as an oscillator producing coherent visible radiation in place of the normal incoherent light from filament lamps or discharge tubes. One of the more readily available gaseous laser systems employs a mixture of helium and neon in which the helium atoms are pumped up to the higher excited level by electron impact in the gas discharge, the latter being induced by a radio-frequency field applied to the gas. This upper 2^3S level of the helium is a longlived metastable state and consequently it rapidly passes on its energy to the $2S$ level of the neon atoms which lies at the same height, and in this way the population of the $2S$ level of the neon atoms becomes much larger than that appropriate to thermal equilibrium and in fact significantly greater than the $2P$ level below it. It is therefore, possible for laser action to take place between these two levels and coherent visible radiation can thus be produced in the manner indicated. *Figure 3.13* has been included to show the essential similarity between lasers and masers, but detailed consideration of lasers as such is not really in the topics covered by this text, although they again afford a very good example of the interesting and invaluable interchange that has taken place between the pure scientist and the applied engineers.

93

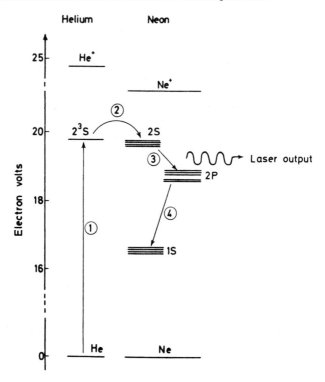

Figure 3.13. Principle of population inversion in helium-neon laser. Initial excitation of helium atoms to the upper 2^3S level is produced by the r.f. discharge. These atoms pass energy to the neon atoms, increasing the population of their 2S level and hence laser action is possible

3.7 Free Radical Studies

As mentioned earlier in this chapter, the two main groups of compounds studied by electron resonance are the transition group complexes on the one hand, and the free radicals of organic chemistry on the other. When the study of free radicals is considered, the situation is inherently more simple than for the transition group complexes, because the unpaired electron is nearly always moving in a highly delocalised molecular orbit and hence interacting very little with any one single atom. As a result there is very little orbital contribution to the g-value, which only shifts very slightly from that for a free spin, and it follows that most of the interesting information must therefore be deduced from other parameters of the spectrum. By far the most informative of these has been the hyperfine splittings which are observed, although the variation of the integrated intensity of the signal under changing conditions can also be of great interest when transient species are being studied.

The early studies on free radicals concentrated on those that had been stabilised by various means, such as the semiquinones already given as examples in Section 3.1, or those that had been produced by a high energy radiation of solid specimens and had then become trapped in the lattice structure. It would probably be fair to say that the main application of such studies as these is that they afford a very powerful method for comparing the detailed predictions of theoretical chemistry with very precise experimental observations. A simple example of this is shown in *Figure 3.14,* where the observed hyperfine structure of the

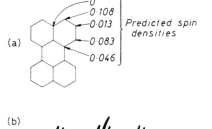

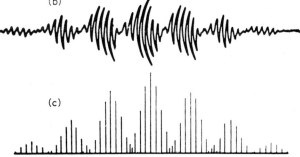

Figure 3.14. Observed and predicted hyperfine patterns for perylene ion: (a) Structure and predicted spin densities; (b) Observed hyperfine splittings; (c) Theoretically predicted lines

perylene ion in solution is shown, and beneath this is given the predicted hyperfine pattern that would be obtained if the spin densities shown in *Figure 3.14 (a)* actually occurred in practice.

95

The values of the spin densities shown were, in fact, those predicted by applying the Hückel molecular orbital theory to this particular molecule, and it can be seen that a very precise check on the accuracy of this theory is available by direct comparison between the observed hyperfine structure and that predicted below it. Nowadays such comparisons can be carried out automatically by computers and in this way the best possible parameters for the molecular structure can be deduced rapidly from the observed electron resonance spectrum. This is a field where experimental techniques and development of theoretical methods have moved hand in hand, and Hückel's original theory has been extended and modified by the introduction of con-figurational interaction and other effects. These can then be checked experimentally, not only against the observed hyperfine structure from the hydrogen atoms, but also from the C^{13} hyperfine splitting con-stants, which are now available from the more sensitive spectrometers.

Although the study of stable free radicals of this type enabled the theory of free radical structure to be developed rapidly and also provided the background techniques for the experimental methods, there is no doubt that the main interest in practice, for free radical studies, is in the investigation of transient radicals which take part as intermediates in various chemical or biological processes. The study of such transient reactions has therefore developed as one of the main interests in such electron resonance studies, and some of the methods which have been devised for this have already been summarised in Section 3.2. A good example of the power of this technique is the study of enzymes by such methods and the work on xanthine oxidase by the sudden freezing technique serves to illustrate this very well. In this particular reaction three different valency changes are produced in the transition group ions associated with the enzyme, while a high concentration of free radicals comes into transient existence at the same time. The kinetics of the valency changes, and their comparison with the free radical concentration, can be plotted simultaneously, as is shown in *Figure 3.15,* the measurements being taken for different time intervals in which the reacting solutions were passing down the reaction tube before deep freezing, as shown in *Figure 3.6 (b).*

It can be seen from the results plotted in *Figure 3.15* that the free radical signal from the flavin in the enzyme rises very quickly during the first 20 ms, then turns over sharply, and finally decays over the next several hundred milliseconds. The signal labelled Mo-δ has a turnover point within the first 20 ms period, whereas that labelled Mo-β does not reach its maximum until about 50 ms. It can also be seen that the one attributed to the iron atom rises noticeably more slowly and does not reach its maximum until about 100 ms. It is obvious therefore, that the various steps in the enzyme activity can be clearly differentiated in this way, and that, for instance, the change of valency in the iron atom

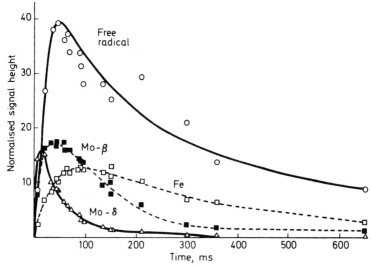

Figure 3.15. Kinetic studies of transient changes in xanthine oxidase. These curves were obtained by the rapid freezing method and the changing concentration of the different metal ions together with that of the specific free radical transient can be clearly seen

occurs much later than the changes in the molybdenum atom, contrary to what had at first been assumed. The ability to plot these valency changes in transition group atoms, at the same time as the transient free radicals occur, illustrates very well one of the specific advantages of electron resonance studies. It may be noted that the identification of these signals can be further confirmed by studying the hyperfine patterns, and in the case of xanthine oxidase, which is the enzyme responsible for milk production, this was actually achieved by feeding cows with artificially enriched isotopes of ^{95}Mo, the hyperfine structure from which could later be identified in the reacting enzymes.

Another field of study in which transient free radicals have been studied in great detail are those produced by photochemical changes, and these studies have introduced another technique which is also proving of great importance in these and other related studies. It was mentioned in the earlier sections on experimental methods that the sensitivity of any electron resonance spectrometer could be increased by incorporating a 'computer of average transients' which stored successive sweeps through the spectrum in a computer or digital memory, and then, after many sweeps had been so recorded, presented the observed spectrum integrated out of background noise. The basic principle behind such a technique is that when successive sweeps through the resonance are fed to the storage memory the different channels of the memory, which correspond to different incremental field values, will have the actual absorption signal component added linearly to them sweep after sweep. On the other hand, the noise which is fed into the channels at the same time is a random effect and will only be summed as the square root of the number of sweeps and as a result after n sweeps the signal to noise ratio will have improved by a factor of $n/\sqrt{n}$ i.e. $n^{\frac{1}{2}}$. Thus, if a signal is buried at a strength ten times below the overall noise of the spectrometer on a single sweep through the resonance line, it is possible to bring this signal up to the noise level by sweeping through the spectrum one hundred times and adding these successively into the store of the computer. It is then possible to further increase this signal strength to ten times above the noise level by sweeping through ten thousand times and storing the sweeps in this way. It is clear, however, that for this technique to work the whole apparatus must be extremely stable, since the conditions of the spectrum must be absolutely repeatable if linear summation of the signal on each sweep is to be obtained.

The argument so far has only been concerned with static signals, but exactly the same argument can be applied to a transient signal if the variation across the channels of the computer storage is now one with time rather than with magnetic field. As a specific example, a case of a photochemical reaction may be taken and the way in which this

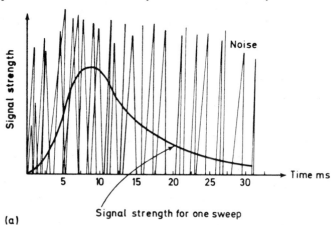

Figure 3.16. Integration of rapidly changing spectrum from noise: (a) Growth and decay of photo-induced signal—below noise level

98

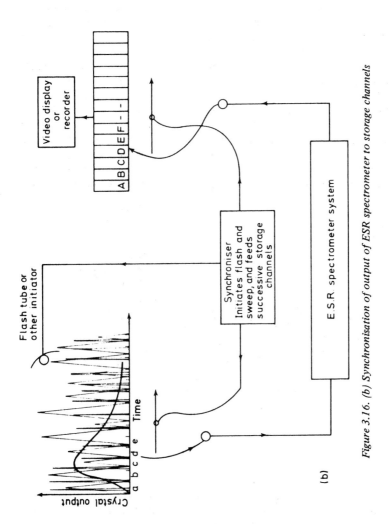

Figure 3.16. (b) Synchronisation of output of ESR spectrometer to storage channels

99

can be followed by using these integrating techniques is illustrated in *Figure 3.16.* where *(a)* represents the growth and decay of a particular photochemically induced species and the time axes can be divided into segments as shown. Thus, any single sweep is initiated by the flash of the phototube which also starts the timing mechanism so that the output from the ESR spectrometer is successively fed to a series of different storage channels, which correspond to different time interval segments as indicated in *Figure 3.16 (b)*. After one single flash and time sweep the signal actually stored in the computer channels will be identical to that of *Figure 3.16 (a)* and hence, if recalled onto the display, would still be completely swamped with noise. On the other hand, this same time-sequence sweep can now be initiated again and the same process repeated. It can, in fact, be initiated and gone through a large number of times, say n, and the signal then stored in the computer channels will be n times that of *Figure 3.16 (a)*, whereas the noise stored in any segment will only be $n^{\frac{1}{2}}$ times the back-

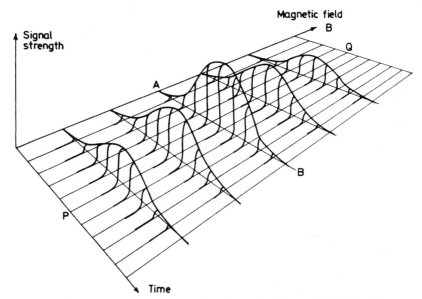

Figure 3.17. Variation of spectrum with time and field strength. This total three-dimensional plot can be built up by plotting successive slices parallel to the time axis at successive fixed magnetic field values

ground of *Figure 3.16 (a)*. It is thus possible in this way to integrate the transient signal variation with time out of its noise background and follow the kinetics of such photochemical reactions in detail.

100

In describing the above method it has been assumed that the field value of the electron resonance spectrometer has remained tuned to the maximum absorption throughout all the observations. It is, of course, possible to study and obtain permanent records of the variation in time of such a rapidly changing signal and also its variation with field. This total map of information is shown diagrammatically in *Figure 3.17* as a three-dimensional perspective, showing how the spectrum will vary with time and also what its particular features, such as hyperfine splittings, may be. At any given instance of time the spectrum might have a profile in its magnetic field variation, as represented by PQ in the diagram, but it would be difficult to trace this particular hyperfine pattern in detail since it would change during the process of recording. Instead, the techniques described above can be employed to trace the time profiles for each magnetic field value in turn, such as represented by the slice AB. Once this information has been obtained and recorded it is then possible to change the setting of the magnetic field and obtain successive time slices and in this way build up the complete pattern of information. Obviously such detailed information as this can be extremely helpful when interpreting transient effects in photochemical or other similar systems. Other specific examples of such applications are considered in the last chapter, where the ability of electron resonance to follow such rapidly changing systems is seen to be one of its considerable advantages.

Chapter Four

Magnetic Resonance at Radio Frequencies

4.1 The Principles of Nuclear Magnetic Resonance

The basic principles of nuclear magnetic resonance have already been discussed briefly in Section 1.6 and, as seen there, the underlying concepts are the same as those of electron resonance, as discussed in the last chapter. The only essential difference is the fact that the spins which are reorientated in the magnetic field are those of nuclei, with their very much smaller magnetic moments. Hence, the energies required for such a transition, and the frequency of the radiation inducing them, are some two thousand times smaller than those corresponding to the electron case. The basic resonance equation for a nuclear magnetic resonance therefore becomes

$$h\nu = g_N . \beta_N . B \qquad (4.1)$$

where β_N is the nuclear magneton, and has the mass of the proton inserted in the expression eh/mc, instead of the mass of electron, and is thus some two thousand times smaller than the Bohr magneton.

If the quantitative figures for the g-value of the proton and the nuclear magneton are inserted into this equation the resonance condition for protons is then given by

$$\nu = 42.6 \, B \text{ MHz} \quad \text{or} \quad B = 2.349 \, \nu \text{ tesla} \qquad (4.2)$$

and, for a typical field of 1 tesla (i.e. 10 000 gauss), the frequency of nuclear proton resonance will be 42.6 MHz. Hence, all the techniques required for experiments of this type will be those associated with the radio-frequency region of the spectrum, such as the balanced bridge circuits incorporating inductances and capacitances, as already illustrated in *Figure 1.9.*

There is, however, an alternative method of detecting the nuclear magnetic resonance absorption, which can be applied very easily in the radio-frequency region, and has come to be employed in a large number of commercial spectrometers. This is known as 'the nuclear induction technique' and relies on the fact that, at resonance, a small radio-frequency field will be induced at right angles to the field of the driving coil. The existence of this perpendicular field can be detected by a coil, suitably placed, which provides a very sensitive method for detecting the resonance condition. The basic principles of this method of nuclear induction are shown in *Figure 4.1.*

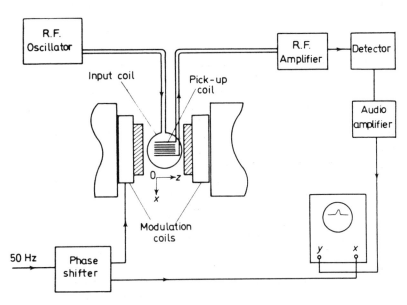

Figure 4.1. NMR spectrometer employing principle of nuclear induction

It can be seen from this diagram that the input coil, which is drawn in the plane of the paper, is fed by a radio-frequency oscillator and has no direct connection to the final output, or pick-up coil. The sample which contains the nuclei to be investigated is placed at the centre of both coils, the pick-up coil having its axis perpendicular to that of the input coil, and also perpendicular to the direction of the applied d.c. magnetic field. This detecting coil feeds any voltage induced within it to a radio-frequency amplifier, the detected

103

output of which is applied to the Y plates of an oscilloscope via an audio amplifier.

At non-resonant values of the applied magnetic field the moments of the nuclei in the specimen will be partially aligned along the direction of the field to produce an overall resultant magnetisation M_Z. The radio-frequency field which exists in the driving coil will not affect this magnetisation to any measurable extent at off-resonance frequency values, and as a result, no induced magnetisation at right angles to the direction of the applied field is produced and no signal is picked up in the detecting coil. If, however, the value of the d.c. magnetic field is adjusted to the resonance value, then the magnetisation, M_Z, will be very markedly affected by the radio-frequency field of the driving coil and will start precessing around the direction of the applied field, at its Larmor frequency. As a result, a rotating component of magnetisation will be produced at right angles to the applied field and this precessional motion will induce a voltage in the detecting coil. This voltage is passed on, via the audio amplifier, and is displayed on the oscilloscope as an absorption line as shown.

As in the case of electron resonance, a small a.c. modulating field can also be applied to the electromagnet so that the total applied field is swept to and fro through the resonance condition. The time base of the oscilloscope is fed in synchronism with this modulation, so that the absorption signal can be displayed, as is usual for a video system. The sensitivity of this rather simple system can be increased by a variety of electronic means, but the details of these are not appropriate in this text. It should be noted, however, that on the practical side, care must be taken to prevent direct pick-up between the input and detecting coils, and, in order to facilitate this, some form of adjustable artificial coupling is usually incorporated so that it can be used to compensate that already present. It may be noted that most modern commercial nuclear resonance spectrometers now employ this induction method rather than the balanced radio-frequency bridge technique.

Apart from the incorporation of more sophisticated electronic circuits over the years, there have been one or two major developments on the experimental techniques side which should be mentioned to illustrate the way in which nuclear resonance has developed. A large amount of the experimental work undertaken in nuclear resonance has been concerned with the high resolution spectra that are observed in solutions, and hence one of the main experimental requirements has been the production of very uniform and homogeneous magnetic fields. Thus, if the maximum amount of information is to be deduced from the high resolution spectra that can be obtained, the inhomogeneity of the magnetic field over the specimen itself must be less than a

104

milligauss. This corresponds to a field homogeneity of 1 part in 10^7 or 10^8 across a specimen of about 4 mm diameter, and normally such extremely high orders of uniformity can only be achieved in magnets with pole face diameters of 20 cm or more. Such magnets have pole faces which have not only been carefully aligned and optically polished, but also usually incorporate small coils attached to the pole faces themselves through which balancing currents can be passed. These thin flat stacks of coil pairs, placed on opposing faces of the magnet, are designed to produce correcting magnetic fields which can be described by different sets of spherical harmonics. In this way the current adjustments in the different coils can be made independent of each other, and the field homogeneity can thus be steadily optimised as the current flowing through each pair in turn is adjusted. In addition to optimising the actual field homogeneity, it is also possible to obtain higher resolution by rapidly rotating the specimen, so that field inhomogeneities over its volume can be averaged out. Thus, if a variation of the magnetic field over the sample is given by ΔB, and the sample is rotated at such a speed that each part is exposed to the variation in a time, t, then the nuclei will see an averaged field, provided

$$t < 2 \pi / \gamma \Delta B \qquad (4.3)$$

where γ is the nuclear gyromagnetic ratio.

For example, if ΔB = 1 mgauss, then this inhomogeneity can be averaged out for a proton resonance if t is less than 1/20s, in other words, provided the specimen tube is spun at a faster rate than 20 rev. s^{-1}, which is quite simple to achieve in practice. It should be noted that the averaging takes place around circular paths, and thus although it averages out the inhomogeneities in the plane perpendicular to the axis of rotation, it does not produce any averaging effect along the vertical axis. On the other hand, it is very much easier to locate the region of maximum homogeneity in the vertical direction, and hence find the point of optimum homogeneity for the spinning technique. This technique will increase the effective resolution by more than a factor of 10 in most cases.

It is also necessary to have a 'time stability' of the magnetic field which is of the same order as its space homogeneity. Thus, if it takes 10s to sweep through a resonance line, it is imperative that the magnitude of the field strength should remain constant to 1 part in 10^8 during this time, if the resolution is to be maintained at this level. Time stabilities of this order of magnitude require highly stabilised power supplies, with automatic correcting circuits, if electromagnets are employed. Permanent magnets have some advantages in this connection, but they do require very precise temperature control to obtain the required constancy of field, and they also have the great drawback that the frequency of the nuclear resonance may have to be changed over a wide range.

In this connection, it should also be pointed out that it is always best to work at as high a value of magnetic field as possible, since this will increase the energy level separation between the spin levels and thus also increase the difference in population between them. This difference in population is the main factor determining the sensitivity of the spectrometer and higher sensitivity will therefore always be obtained at the higher values of magnetic field and radio frequency. Moreover, differentiation between the different types of splitting and shift that occur in the spectra can also be best obtained at the higher values of field and frequency, as discussed in the next section. Typical values employed in modern nuclear resonance spectrometers are frequencies of 100 MHz which require a field of about 2.3 tesla (23 000 gauss) for a proton resonance. Such fields represent more or less the limit, so far as conventional electromagnets or permanent magnets are concerned, although it is possible to exceed these, and work at higher fields and frequencies, if superconducting magnets are employed.

4.2 N.M.R. Spectra from Solids and Liquids

Most of the early studies on nuclear resonance were undertaken on solids, and often on single crystals. It was found that a considerable amount of information could be deduced by studying the line shapes and splittings of the proton resonance signals obtained from such single crystals. The general principles governing the theory of linewidth and interaction in nuclear resonance follows very closely along the lines already discussed for electron resonance. Thus, the most common interaction that will be present will be the direct dipole–dipole interactions between neighbouring nuclei, and in a solid this will follow the normal $(3 \cos^2 \theta - 1)$ variation, which has already been discussed in some detail for electron resonance. One of the first striking examples of such an interaction came from the studies of a single crystal of gypsum, i.e. $CaSO_4 . 2H_2O$. This gives a pair of doublet lines, the splitting between them depending on the orientation of the crystal. This splitting can be explained by assuming that the two protons of each water molecule act together as a single group, and thus each of the two molecules produces a doublet, the separation between them depending on the angle which the direction of the inter-proton axis makes with the direction of the applied magnetic field. Although an analysis of such splittings and linewidths can give direct information on the structure of the molecules, such lines obtained from the solid state are usually rather broad, and accurate quantitative measurements are therefore rather difficult to make.

On the other hand, when solutions are studied very narrow resonance lines are obtained, hence very precise and accurate measurements can be made and correspondingly detailed information drawn on the structure of the molecule. The presence of the very narrow lines in liquids or solutions is due to the motional narrowing that occurs for the nuclear spins, in exactly the same way as already discussed for the electron spins. Thus, the rapid tumbling motion of the molecules in the liquid state will cause all the nuclei to pass rapidly through the whole range of values, corresponding to the $(3 \cos^2\theta - 1)$ variation, in a time short compared with the inverse of the frequency, and hence the effective field they experience will be averaged to zero. Before discussing the high resolution spectra that can thus be obtained in a liquid, intermediate states between the solid and liquid are briefly considered since they can also be very effectively studied by the nuclear resonance technique.

Thus, the onset of internal molecular rotation, within a solid, can be quite dramatically followed by observing the sudden decrease in the width of the nuclear resonance absorption line, since this motional narrowing will set in immediately the particular group under study starts rotating, even though the specimen as a whole still remains solid. The direct link between rotation and the narrowing of resonance lines was demonstrated very directly by a very elegant set of experiments in which a crystal of sodium chloride was spun at speeds of up to 50 000 rev. min^{-1} by an air-driven rotor. If the axis of rotation is adjusted to make an angle of 54°44′ with the direction of the magnetic field, the angle for which $(3 \cos^2 \theta - 1) = 0$, the dipole–dipole linewidth of the ^{23}Na nuclear resonance, which is about 1 gauss in the static crystal, can be effectively reduced to zero by the rotational narrowing. The absolute intensity of the signal is not entirely lost in such a rotational narrowing process but reappears as harmonic side bands, well separated from the main resonance line. These harmonics will only be clearly observed if all the molecules are rotating at the same speed, and will be spread over a wide range of frequencies and be undetectable in the random motion of a liquid sample.

It would be fair to say, however, that the major application of

107

nuclear resonance has been in the high resolution spectra that are observed from liquids or solutions, and the various additional features which can be observed in these are now discussed in some detail.

4.3 The Chemical Shift and Spin—Spin Coupling

As remarked earlier, there are two great advantages of studying spectra in solution. One is that high resolution is available because of the motional narrowing, and the other is that all the interactions which vary with angle having been averaged to zero, the spectra are relatively straightforward to analyse since only isotropic interactions are present. There are two major types of isotropic interaction which occur in the liquid state and which will produce a shift, or splitting of a single resonance line.

4.3.1 The Chemical Shift

The first of these interactions is termed the 'chemical shift', which is the name given to the fact that the same nucleus will not always have its resonance absorption line at the same value of magnetic field for a given value of the applied radio frequency. This fact, and the reasons for it, can probably be best illustrated by taking a specific example, such as ethyl alcohol. This has the structural formula:

$$
\begin{array}{ccc}
 & H & H \\
 & | & | \\
H-O- & C- & C-H \\
 & | & | \\
 & H & H
\end{array}
$$

The vast majority of carbon and oxygen atoms in these molecules have no nuclear spins or magnetic moments, since they are nearly all ^{12}C or ^{16}O nuclei. Hence, resonance is only to be expected from the hydrogen nuclei, i.e. the protons, and a low resolution nuclear resonance spectrometer will display a single absorption line at a magnetic field strength corresponding to the resonance frequency of the proton, each of the six protons in the molecule contributing to its intensity. If a somewhat greater resolution is employed, however, the single absorption line is found to split into three components, as illustrated in *Figure 4.2 (a)*. Moreover, these three components are found to have intensity ratios of approximately 3:2:1. This fact suggests very strongly that they must be arising from the three different kinds of proton which are to be found in the alcohol molecule, i.e. the larger line corresponds to the resonance of the three protons of the CH_3 group, while the central line corresponds to the two from the CH_2 group, and the smaller line corresponds to the one proton of the OH group.

The fact that these three resonances are now observed for the same incident radio frequency shows that the different types of protons must be residing in slightly different values of magnetic field strength. Since the same external magnetic field is applied across all the protons this suggests there must be different internal fields arising within the molecule, which are responsible for these small splittings. These internal magnetic field contributions are, in fact, produced by the different effects of the diamagnetic shielding formed by the outer electrons of the chemical bond, which are slightly different for the three chemical groups concerned.

Thus, the electron cloud distribution around the protons in the CH_3 group will be slightly different from that around the CH_2 group, and the diamagnetic effect in the resultant shielding from the applied magnetic field will be slightly different in the two cases. The fact that this shift, or splitting, is due to such a mechanism is confirmed by the dependence of the magnitude of the splitting on the value of the applied d.c. field. Thus the higher the value of the main magnetic field, the larger is the splitting found between the three component lines, confirming that it is indeed a diamagnetic effect, which is linearly dependent on the value of the applied magnetic field.

Such a shift or splitting will be characteristic of the particular chemical group in which the proton resides and for this reason it has come to be called the 'chemical shift'. It can be placed on a quantitative basis by defining a parameter, δ, in terms of the following expression:

$$\delta = \frac{B_{obs} - B_{ref}}{B_{ref}} \cdot 10^6 \qquad (4.4)$$

B_{obs} is the value of the magnetic field at which the particular resonance is observed, while B_{ref} is the magnetic field strength at which the signal from a reference group occurs, for the same radio frequency. Provided that the same substance under the same conditions is always taken as a reference, it is immaterial which group is actually employed as such a reference. The two protons most commonly used for this purpose are the protons in water, when aqueous solutions are studied, and the protons in benzene when organic compounds are under

109

investigation. The factor of 10^6 is added in the equation above because the shifts observed are extremely small and a multiplying factor of this type then produces reasonable values to be quoted for the actual shifts themselves.

A list of some of the chemical shifts that are observed for protons in different chemical groups is given in *Table 4.1,* and it can be seen that the magnitudes quoted there are of a positive and negative sign, which only indicates that the diamagnetic shielding for some groups is greater than that for the protons in water, whereas for others it is less.

Table 4.1
CHEMICAL SHIFTS FOR PROTONS IN DIFFERENT CHEMICAL GROUPS
(WATER TAKEN AS REFERENCE)

Group	δ	Group	δ
$CH_3 - C \equiv$	+3.8	$CH_3 - 0 -$	+1.5
NH_2 (alkylamine)	+3.7	H_2O	0
CH_2 (cyclic)	+3.5	$CH_2 =$	−0.4
$CH_3 - C =$	+3.4	OH	−0.5
CH_3 ⟨⟩	+3.3	$C = CH - C$	−0.6
$CH_3 - \overset{\vert}{C} =$	+3.2	⟨⟩	−2.0
$CH_3 - N$	+2.5	NH_3 (amide)	−2.8
$HC \equiv$	+2.3	$C - CHO$	−4.7
$C - CH_2 - X$	+2.0	COOH	−6.5

It is now evident that the accurate measurement of the magnetic field strength at which any given proton resonance occurs, and the comparison of this value with that obtained from a reference compound under the same conditions, can be used to determine the chemical shift of the unknown proton and hence to identify the chemical group in which it resides. Such a measurement on its own provides a very powerful analytical tool, and although the specific cases quoted have been those of compounds containing protons, exactly the same type of chemical shift will also occur for other nuclei, and these can be used in the same analytical way as for the protons. The sensitivity of nuclear resonance spectrometers is now sufficiently high that signals can be obtained from ^{13}C, as it occurs in its natural abundance in organic compounds, and the chemical shifts observed for this provide a powerful additional analytical tool.

4.3.2 Spin–Spin Coupling

If the nuclear resonance spectrum of ethyl alcohol is studied under still higher resolution, further splittings of the three resonance lines

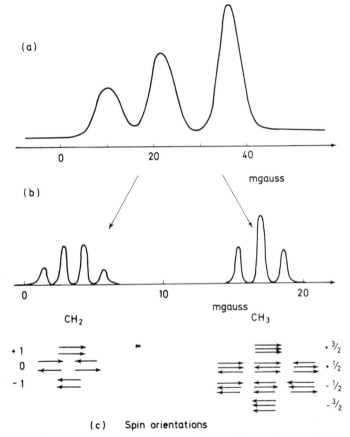

Figure 4.2. NMR spectrum of ethyl alcohol: (a) Under medium resolution; (b) Under higher resolution; (c) Spin orientations of CH$_2$ and CH$_3$ groups of protons, showing combined effects

will be obtained, as shown in *Figure 4.2 (b)*. It can be seen that the CH$_3$ resonance line has now been split into three different components,

111

with the centre component twice as intense as the two outer ones; while the CH_2 line has been split into four components with the intensity ratios 1:3:3:1. This further splitting cannot be due to any chemical difference between these identical nuclei, but is in fact due to the coupling between the spins of protons in different groups on neighbouring sites.

Thus, all three protons in a CH_3 group are not only affected by the external magnetic field, and the diamagnetic shielding characteristic of the methyl group, but they are also quite close to the protons of the CH_2 group, and these will act as small magnetic dipoles to produce a small additional field at the site of the protons of the CH_3 group. There are various possible ways in which the two protons of the CH_2 group can be arranged amongst themselves. Their nuclear spins and magnetic moments may both be pointing in the direction of the applied magnetic field and in this case they will produce an incremental effect in the direction of the field. It is possible, however, for both of these spins to be revised, as indicated in *Figure 4.2 (c)*. In this case the same magnitude of the incremental field will be produced at the methyl protons but it will now be in opposition to the main applied field. It is, moreover, also possible for the two spins and magnetic moments of the CH_2 protons to be pointing in opposite directions, thus cancelling out, and not producing an additional field at the site of the CH_3 group. It is also evident that there are two ways in which this null result can be achieved, whereas there is only one way in which either of the two other extreme effects are produced. It therefore follows that, if the statistical average is taken of all the ethyl alcohol molecules, there will be twice as many with their CH_2 protons cancelling, than with their proton spins lined up together, either in the direction of the applied magnetic field or against it. This explains why the additional splitting exists in the CH_3 proton resonance line, and also why the centre line, which corresponds to the unshifted resonance, twice as intense as the two other lines.

This argument can now be applied in reverse to consider the effect of the CH_3 protons on the resonance produced by the protons from the CH_2 group. In this case there are four different ways in which the three proton spins and magnetic moments of the CH_3 group can be orientated. They can all be lined up in the direction of the field to produce the maximum shift in that direction, or alternatively, all be lined up against the field to produce a maximum decrease. The two other possibilities correspond to two of the three protons cancelling and leaving only one unpaired spin and magnetic moment to produce a net effect. This single unpaired moment can,

112

of course, be either acting with, or against, the direction of the field and hence there will be two alternative small additional fields corresponding to these two cases. Moreover, as will be seen from the figure, there are three different ways in which each of these two intermediate effects can be produced, and this again explains why the two centre lines of the additional splitting of the CH_2 resonance are three times as intense as the two outer lines. It is clear, therefore, that this additional splitting, due to the coupling between neighbouring groups, gives a large amount of additional information and enables, not only different chemical groups to be specifically identified from the chemical shift, but also their position, relative to each other, to be accurately determined.

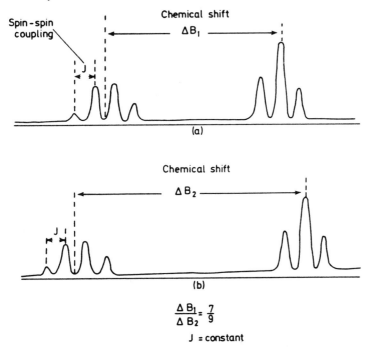

Figure 4.3. Resonance from ethyl alcohol at different field strengths: (a) at 0.7 tesla; (b) at 0.9 tesla;

On the basis of the above argument, one would expect the single small line of the OH proton to be split into a triplet, with an intensity ratio of 1:2:1 due to its proximity to the CH_2 group; although the magnitude of the splitting will not be quite as great as

113

that of the CH_3 group, since the OH proton is somewhat further away from the CH_2 group than are the protons of the methyl group. However, this splitting of the OH resonance line is not normally observed, due to the rapid exchange that the OH protons are undergoing with other similar protons in the liquid. If the rate of this exchange is faster than the frequency corresponding to the spin–spin splitting, then the splitting will be averaged to zero in exactly the same way as the dipolar splitting was averaged by the motional narrowing considered before. In very pure specimens, where the exchange rate is slow, it is possible to observe the triplet splitting of the OH proton. The sudden disappearance of a spin–spin splitting can often be used to

Table 4.2
CHEMICAL SHIFTS AND SPIN–SPIN COUPLING CONSTANTS OF ^{13}C
(BENZENE TAKEN AS REFERENCE FOR CHEMICAL SHIFTS)

Compound	δ p.p.m.	J_{CH} H_z
$CH_3 I$	150	151
$CH_3 Br$	124	153
$CH Br_3$	115	208
$CH_2 Br_2$	110	185
$C_6 H_{12}$	100	140
$CH_3 OH$	80	144
$CH_2 Cl_2$	75	162
$CH_3 NO_2$	58	137
$CH Cl_3$	50	193
$C_6 H_6$	0	159
HCOOH	−38	218
$CH_3 CHO$	−72	174

indicate the onset of exchange between neighbouring molecules, and this can then be employed as a powerful probe of the internal state of the specimen under investigation.

In the case of ethyl alcohol it has been relatively easy to identify

114

and explain these two different types of splitting, but in a more complicated molecule the chemical shift and spin—spin coupling splittings may well overlap, and the correct identification may then prove somewhat difficult. Fortunately there is a very simple method by which they can be differentiated and this arises from the fact that the chemical shift varies linearly with the applied magnetic field, as previously discussed, whereas the spin—spin coupling, which is determined entirely by the spacing between the chemical groups in the molecule, will be independent of the applied magnetic field. Thus, if a high resolution spectrum of ethyl alcohol is taken at two different radio frequencies the spectra will be as shown in *Figure 4.3 (a) and (b)*. It can be seen from these that the chemical shift increases linearly with the magnitude of the applied field, whereas the splitting due to the spin—spin coupling remains the same at both frequencies. It is for this reason that measurements at two different frequencies are often extremely valuable when complex spectra have to be analysed in detail. Since the magnitude of the spin—spin coupling constant is independent of magnetic field it can be quoted directly in Hz, and typical values for both the chemical shift and spin—spin coupling constants of ^{13}C in various compounds, are given in *Table 4.2* where, in this case, the reference for the chemical shift has been taken as the ^{13}C in benzene.

4.4 Exchange and Motional Effects

In the last section we were concentrating on the various splittings which are produced in the spectra by different chemical groups, and their interactions with one another. In addition to this information, which is basically of an analytical or structural kind, information can be obtained on the kinetics of actions taking place between molecules or within the molecule itself. Such information can be deduced from the change in line widths, or the averaging out of splittings, as previously mentioned when the example of the OH resonance of ethyl alcohol was considered. Thus the expected triplet structure of this is often averaged to a single line by the rapid exchange of the OH proton between alcohol molecules and neighbouring water molecules.

   In general terms we may state that if the resonance from a proton in molecule X is normally observed at a magnetic field strength, B_X and another absorption is obtained from a proton in molecule, Y, at field strength B_Y, then, while these two protons remain firmly attached to their two different molecules, both of the resonance lines will be observed independently with intensities corresponding to the number of protons in the two molecules concerned. If, however, these two protons exchange rapidly, at a frequency which is higher than that corresponding to the difference in energy between the two spectral lines, then both protons will only see an average field and the resonance will be observed at some value between B_x and B_y, the actual value depending on the proportion of protons of type X present to those of type Y.

This exchange-averaging process may be a little more difficult to envisage than the averaging produced by the random tumbling motion in a liquid, but a simple analogy may help to illustrate the effect. Thus, one can imagine a ball being thrown between two players, and being observed by a photographer. Then, as long as the speed of the camera shutter with which he is taking pictures is much faster than the rate at which the ball is thrown between the two players, each photograph taken will show the ball is either with one player or with the other. If, however, the two players are throwing the

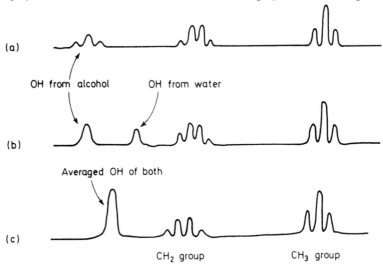

Figure 4.4. Spectra from ethyl alcohol and water mixtures showing exchange and averaging effects: (a) From very pure alcohol showing triplet splitting of OH proton; (b) From mixture with 5% water showing separate OH proton resonances from the alcohol and water molecules; (c) From mixture with 30% water showing complete averaging of OH resonance

ball between themselves at a rate which is much faster than that of the camera shutter speed, then only a blur will be produced on the photograph, the position of the ball having been averaged in the process.

116

It has already been mentioned that the OH group of ethyl alcohol serves as a good example of this, and *Figure 4.4 (a)* shows the spectrum that is observed from very pure ethyl alcohol, in which there are no other molecules containing OH groups with which these protons can exchange, and the triplet structure on the OH resonance is therefore observed. If, however, a small amount of water is present, then exchange can take place between the protons on the OH group of the ethyl alcohol and those on the water molecules, and the triplet on the OH resonance of the alcohol is averaged to a single line, while a second OH resonance from the proton of the water molecules is also observed, as seen in *Figure 4.4 (b)*. The splitting between these two OH resonances is 32 Hz. If this exchange is fast enough, the two OH resonances will be averaged to produce a single line at an intermediate resonance value, as clearly seen in *Figure 4.4 (c)*, the large intensity coming from the larger number of water molecules now present. It may be noted that this change from the two spectral lines to an averaged single line occurs at water concentrations of 25%, and it can therefore be deduced quite specifically that, at this concentration, the exchange of protons between the two molecules is taking place in the time of 0.015s, since the mean lifetime in a given state is related to the splitting, which has been averaged, by the equation

$$\tau = \sqrt{2} \, (\pi \, \delta v) \qquad (4.5)$$

where δv is the splitting that is averaged out.

Averaging by rapid exchange is not the only process which will reduce a splitting of the spectrum, however, and one other very general type of motion which can be studied in this way is that of internal molecular rotation which can take place inside a solid. A good example of this is the spectrum obtained by dimethyl formamide, which has the structural formula:

It is clear that the single proton attached to the CO group will give a resonance well displaced from the protons of the two methyl groups; as is seen to be the case in *Figure 4.5*. The two methyl groups themselves are not in exactly equivalent positions, however, since, if the molecule is rigid, one of these will be closer to the proton and the other to the oxygen atom, and hence a slightly different chemical shift would be expected for the two groups. This can be seen at the top of *Figure 4.5* where the 10 Hz splitting shown between the two methyl group absorption lines is produced by this slightly different environment. If, however, the dimethylformamide, which is solid at ordinary temperatures, is gradually heated then the spectra are seen to change, as shown in the figure.

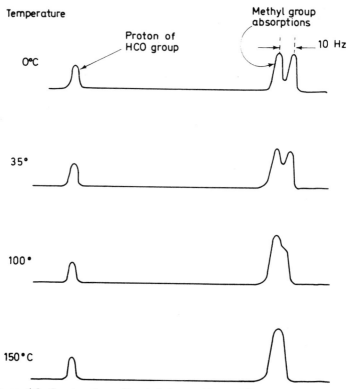

Figure 4.5. Spectra from dimethyl formamide at different temperatures. At higher temperatures the slight difference between the two CH₃ resonances is averaged out by molecular rotation

Obviously the splitting between the two methyl groups becomes successively smaller and at a temperature of 150° C the doublet has coalesced into a single absorption line. This is due to the fact that at this temperature the two groups start rotating about the CN bond and, as a result, average their environments so that when the frequency of rotation is faster than the 10 Hz splitting, a time-averaged resonance position is obtained for both groups.

There are many other types of motion that can be studied in the same way as discussed for proton exchange and internal molecular rotation, and these include such processes as diffusion and hindered rotation, as well as others more specific to particular molecules.

4.5 Nuclear Resonance in Metals

It was noticed earlier that the nuclear resonance lines observed from solids normally have quite a large width, since the dipole–dipole interaction has not been averaged out, but there is one striking exception to this general case and that is nuclear resonance which is observed for metals. In this case narrow absorption lines very similar to those observed in a liquid state are obtained, due to the fact that the metals contain large numbers of free conduction electrons, which set up fluctuations in the magnetic fields at the nuclei and hence average out the anisotropic contributions, in a very similar way to that occurring in the liquid state. A very considerable study has been undertaken on nuclear resonance in metals to follow up these points and in relating theory to experiment it is of course necessary to employ Fermi statistics, since the conduction electrons will obey this form of statistics rather than the normal classical Maxwell–Boltzmann statistics. This means that only those electrons which are near the Fermi surface will contribute to the magnetic effects, since the others well below this energy will have their spins paired.

It is possible to make a reasonable estimate of the time taken by a conduction electron when jumping from one atom to another, in terms of its velocity at the Fermi surface, and the length of the unit cell. This correlation time is found to be of the order of 10^{-16} s and it is clear that this is very much shorter than the inverse of any nuclear resonance frequency, which will be of the order of 10^{-7} s. Hence an averaging effect will certainly be produced by the motion of such conduction electrons and this explains the presence of the narrow lines in the metal.

When the precise position of the resonance line is observed it is found that there is not only a narrowing of the lines, but also what appears to be a chemical shift associated with the nuclear resonance in the metal. This shift of the resonance, when in the metallic state, can also be explained in terms of the presence of the conduction electrons which will produce a net additional magnetic field at the site of the nuclei. In order to determine the average effect of these electrons it is necessary to consider the net magnetisation which they will produce, by applying Fermi–Dirac statistics to their energy

119

states. It is then possible to show that the shift in resonance will be given by the equation

$$B_N = B_O (1 + K) \tag{4.6}$$

where the constant K is termed the Knight shift after its discoverer. This can in turn be related directly to the Fermi energy level of the metal in question by the equation

$$K = 4\pi\beta^2 .p. (P_F/E_F) \tag{4.7}$$

where E_F is the value of the Fermi energy level, and p is the number of valence electrons per atom, while P_F is the constant which relates the wave function of the electron at the Fermi level to the nucleus in question. This is rather an important relation since it shows that it is possible to obtain accurate quantitative estimates of the coupling between the conduction electrons and the nucleus, by a quantitative measure of the Knight shift of the nuclear resonance line. Such information helps to build up the picture of the energy band structure of the conduction electrons themselves, and nuclear resonance measurements on metals have provided a very significant tool in the elucidation of the energy level structure within them. It might be noted in this connection that the resonance frequency observed from nuclei in a liquid metal is almost exactly the same as that observed from nuclei in the solid metal, and the small difference can be accounted for by the discontinuity in density which occurs at the melting point. The fact that there is no real difference between the two frequencies is striking confirmation of the fact that the electron energy states must be very similar in the solid and liquid phases for metallic systems.

Another particular case which is of interest is the nuclear resonance which is observed in metals which are ferromagnetic. In such cases there will be very large internal magnetic fields present within the specimen and these will of course completely alter the resonance condition. The magnitude of these internal fields can be deduced from measurements of the Mössbauer effect, where the shift in the frequency of emitted γ-rays can be used to determined the magnitude of the fields at the nucleus. The precise magnitude of these fields can then be checked by nuclear resonance measurements. Thus, for example, Mössbauer measurements on [57]Fe indicated that there must be an internal magnetic field of about 33 tesla (330 000 gauss). This has been confirmed by the direct observation of the nuclear resonance of the [57]Fe nuclei, at a frequency of 45 MHz, the resonance condition being provided by the internal field of the ferromagnetic rather than any field which is applied externally.

120

4.6 Pulse Methods

So far in all our considerations we have been assuming a steady state for the nuclear resonance condition. Thus, the radio-frequency signal has been assumed as applied continuously as a sine wave, while the d.c. magnetic field is more or less constant all the time, just being slowly moved to and fro through resonance, in order to trace out the absorption line spectrum. It is, however, possible to study nuclear resonance and induction by using pulse techniques, rather than continuous-wave methods. These pulse techniques have become of increasing importance during recent years since automatic methods of storing and analysing the resultant signals now enable all the fundamental information contained within the normal absorption line to be obtained very quickly. The initial pulse and echo methods were introduced very early in the history of nuclear resonance, however, and the basic ideas behind them can be briefly summarised as follows.

A strong pulse of resonance radio frequency is applied to the nuclei already present in the external magnetic field and the duration of such a pulse can be adjusted to orientate the nuclear moments so that they precess to have their axes in a plane perpendicular to the direction of the applied magnetic field. When this pulse is switched off the moments will precess freely about the direction of the main field and will induce a voltage in the receiving coil as they do so. This voltage will decay, and the rate of decay will be determined by the rate at which the nuclear moments lose phase memory and become randomly orientated in the perpendicular plane. At the same time as losing phase in the perpendicular plane, they will start to precess inwards, towards the direction of the applied magnetic field, and the rate of this precession is determined by the spin–lattice relaxation time, T_1, and this may be much longer than the loss of phase memory in the perpendicular plane, which is governed by the spin–spin relaxation time, T_2. It is obvious that, in principle, a study of the rate of decay of this induced nuclear magnetisation can give information on both T_1 and T_2 for the line under question.

Such information can, in fact, be more directly determined by using two such driving pulses and producing what is known as 'spin echoes'. Two short radio-frequency pulses, each of duration τ_w are separated by a time, t_1, such that $\tau_w < t_1 < T_1$ or T_2. It is found that at a further time, t_1, after the second pulse, the nuclei *emit* a pulse of radiation, this occurring in the complete absence of any input radio frequency. The production of this echo pulse can be explained by the constructive interference that takes place, at this particular instant, between the rotating components of magnetisation produced by the

121

two input pulses. Thus, 'during the first pulse, all the vectors representing the nuclear moments will orientate into a plane perpendicular to the direction of the d.c. field. Then, during the time between the two pulses, these vectors will continue rotating in this plane but in random directions reaching an isotropic distribution before the advent of the second pulse, which may be considered as turning the magnetisation through π. In the time following the removal of the second pulse the vectors again rotate in a perpendicular plane, but now in the opposite direction to the first case, so that they are, in fact, unwinding their loss of phase. Therefore, at a time, t_1, later they are all exactly in phase again, and a resultant voltage is induced in the receiver coil. The spins will also be losing orientation due to the spin—lattice relaxation time, however, and hence the spin—lattice relaxation time can be obtained by plotting the variation of the amplitude of these stimulated echoes against the time between the initiating pulses. This simple echo technique can be modified in a large number of ways, by the introduction of third pulses, or by using pulses which correspond to 180° phase changes.

In general terms, it may be said that the basic principle of pulsed nuclear magnetic resonance spectroscopy is to obtain information by broad-band excitation of the nuclear spins with radio-frequency pulses of various durations, and then study the shape of the decaying nuclear induction signal which follows afterwards. These decay curves will, in principle, contain all the information that is present in the normal steady state nuclear magnetic resonance spectrum, and they do in fact represent the Fourier transforms of the resonance line, plotted on a frequency axis. The method of detailed analysis, and a consideration of the various techniques that are now possible using these pulse methods, is beyond the scope of an introductory volume of this nature, but, among other things, the technique can be applied very successfully to solids, where a kind of artificial line narrowing by spin rotation can be produced. To achieve this one employs pulse programmes, which rotate the spin system through an angle of 54.7°, instead of actually rotating the sample with its axis making this angle to the field, as was done in the original high resolution experiments on solids. The increasing popularity of these pulse techniques is due to the advent of computers which can now be linked to the spectrometer to carry out fast Fourier analysis of the results, and hence present the information in a similar way that would have been obtained from the standard absorption spectrum. Results can thus often be obtained much more rapidly than in the earlier steady-state systems.

4.7 Nuclear Precession in Low Field Strengths—Measurement of the Earth's Field

One specific application of the pulse methods that should be mentioned is the way in which these precessional methods have been applied to the measurement of very small magnetic fields, and in particular to accurate measurements of the earth's field itself. The principle of this method can be explained briefly as follows.

The sample containing protons, such as a bottle of water, is placed inside a large single coil and initially a strong magnetising field is applied by this coil to produce a field of some 0.01 tesla (100 gauss), in a direction approximately at right angles to the earth's field (or other field to be measured). As a result nuclear alignment occurs along the direction of this applied field and a net nuclear magnetisation is set up in this direction. This initial polarising field is then switched off very rapidly so that the induced nuclear magnetisation has no time to reorient. The only magnetic field now left is that of the earth itself and the nuclear magnetisation therefore begins to precess around the direction of the earth's field at a frequency given by the normal Larmor, or nuclear resonance, equation, which is about 2 kHz for the earth's field of 5×10^{-5} tesla. This precession dies away in a relaxation time corresponding to the protons in the liquid, but this may be of the order of several seconds, and during this time the rotating nuclear magnetisation will induce a radio frequency signal in the coil. The decaying sine wave can then be fed to an audio amplifier, and either displayed on an oscilloscope or used to activate a frequency counter. An accurate measurement of the frequency of precession can therefore be easily made, and this immediately gives the magnitude of the total component of the earth's magnetic field at the point of the specimen. Moreover, the inhomogeniety of the field is also given directly by the decay time of the precessing signal, as can be seen in *Figure 4.6*. The signal shows the trace obtained from nuclei precessing in the earth's field at a point where this is very homogeneous, and it is seen that a total relaxation time of 10s is then obtained. The second signal is for the identical specimen, but with a razor blade placed a few centimetres away. This produces sufficient inhomogeneity in the field to cause the protons to loose phase coherence much more rapidly, and as a result the decaying signal lasts for a significantly shorter time.

It will be appreciated that this technique not only gives a very direct method for obtaining very accurate measurements of the earth's field, but it is also an extremely simple piece of equipment.

123

It has therefore had many applications wherever rapid measurement of the earth's field is required, such as in geological prospecting for oil and archaeological exploration for buried material. In nearly all of

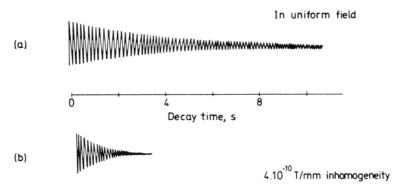

Figure 4.6. Nuclear precession signals obtained in earth's magnetic field: (a) in a very uniform field; (b) in an inhomogeneous field of 4 microgauss/mm

these cases it is the *variation* in the earth's magnetic field strength rather than its absolute magnitude which is of importance. An extremely simple instrument can then be made to detect and measure field variations, which takes the form of two specimen bottles, with associated coils, mounted a metre or so apart on a wooden rod. Signals from these two different groups of protons are then mixed, and their beat frequency detected, and this beat frequency can be fed to a meter and to a loudspeaker. The onset of any significant changes in the earth's field strength will then produce a gradient between the field value at the two specimens and hence slightly different precession frequencies will be generated, which result in a very clear audible beat note from the detecting loud speaker. Such an instrument can be built for a few pounds and yet it can measure changes in the earth's field strength of much less than a milligauss and is now proving quite a popular adjunct to amateur divers in search of sunken ships!

4.8 Double Resonance Techniques

This chapter has been entirely concerned with resonance phenomena linked with the nuclei of atoms, whereas the last chapter was entirely concerned with resonance phenomena linked with the unpaired electrons in atoms, or molecules, and so far there has been no suggestion that these two resonance phenomena could be studied together in the same specimen. The interesting question does now arise, however, as to what exactly would happen if both nuclear magnetic resonance and electron spin resonance are carried out simultaneously on the same specimen. The possibility of such double resonance experiments, and the information that can be deduced from them, has opened up quite a fascinating field of study in recent years. There are in fact various ways in which such double resonance experiments may be conducted. The first of these, known as the Overhauser effect, enables the high sensitivity of the electron resonance technique to be made available in nuclear magnetic resonance studies, whereas a second form which followed a few years later and has come to be known as ENDOR (electron nuclear double resonance), allows the opposite exchange of attributes, in that the high resolution associated with nuclear resonance can then be obtained in an electron resonance spectrum. In both of these methods the phenomenon of saturation, as discussed in Section 3.5, plays a very important part, and in both cases there is some mechanism for coupling between the electron and nuclear spins, which may be either as a direct hyperfine interaction or of a more general nature.

4.8.1 The Overhauser Effect

In the case of the Overhauser effect, a large microwave signal is employed to saturate the electron resonance absorption of unpaired electrons which are loosely coupled to the nuclei under investigation. The change in the normal population distribution of the electron levels, produced by the saturating input power, couples to the nuclear spins and produces a very marked change in their normal energy level population as well. As a result of this it is possible to enhance the nuclear resonance absorption, and in this way very much higher sensitivity is achieved in nuclear resonance studies. The power of this technique is very vividly demonstrated in *Figure 4.7*, which shows successive sweeps through one of the proton resonance lines of dimethylformamide, observed initially in the absence of the saturating microwave power and then in its presence. The four small humps on the left are the successive traces of signal, with no microwave pumping, while the four large inverted signals are those obtained in successive sweeps when the microwave power is applied. The coupling between the unpaired electrons and the nuclei is provided by simply dissolving some organic free radical into the liquid, the nuclear resonance of which is to be studied, and then saturating the electron resonance signals of the free radical. Theoretically all that need to be done in this technique is to apply the microwave radiation at the correct frequency and there is no need to detect or display it. However, it is often better to build a complete resonance spectrometer in order to check that the electron resonance is being obtained and saturated. The actual enhancement obtained in a given situation depends both on radical concentration and the incident microwave power, but in practice it is easy to obtain an improvement in sensitivity by at least a factor of 100.

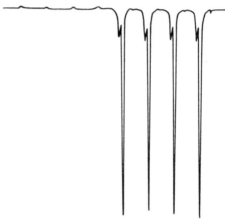

Figure 4.7. Enhancement of NMR signal by Over-hauser effect. Signals observed from dimethyl formamide: left, in the absence of saturating ESR power; right, with it

4.8.2 The ENDOR Technique

In the ENDOR technique high power microwave radiation is again used to saturate the electron resonance transition, but it is this transition itself which is detected and displayed and not the nuclear resonance signal. The nuclear resonance power, in this case, is used to desaturate the electron resonance for certain specific frequencies, and thus in this way locate fine structure present within the main electron resonance spectrum. The method can probably be best illustrated by taking a specific example, and the case of an electron interacting with one proton may serve this purpose.

In this case each of the original single electron resonance levels will be split into two by the interaction with the nuclear spin of the proton, as shown in *Figure 4.8*, and this energy of hyperfine interaction

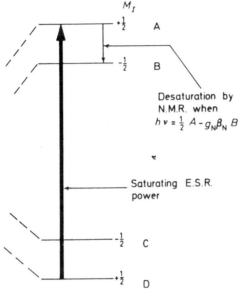

Figure 4.8. Basic principle of ENDOR technique. High power microwave power saturates the electron resonance between A and D and this is then desaturated by the applied NMR frequency

can be denoted by $\frac{1}{4}A$. In addition to this normal hyperfine interaction between the magnetic moments of the unpaired electron and the proton, there is also a small direct effect of the externally applied magnetic field, B, on the nuclear magnetic moment itself. This is usually much smaller than the hyperfine interaction in normal electron resonance spectra, but it will produce a small shift in the four component levels, as indicated in *Figure 4.8*. The actual energy of these four levels can therefore be tabulated as:

Level	Energy
A	$\frac{1}{2}g\beta B + \frac{1}{4}A - \frac{1}{2}g_N\beta_N B$
B	$\frac{1}{2}g\beta B - \frac{1}{4}A + \frac{1}{2}g_N\beta_N B$
C	$-\frac{1}{2}g\beta B + \frac{1}{4}A + \frac{1}{2}g_N\beta_N B$
D	$-\frac{1}{2}g\beta B - \frac{1}{4}A - \frac{1}{2}g_N\beta_N B$

It can be seen from these expressions, and also from the energy-level

127

spacing in the figure, that the difference in energy between levels A and B is now not quite the same as that between C and D.

In the ENDOR technique high power microwave frequency is applied to saturate one of the electron resonance transitions, such as that between A and D in *Figure 4.8*, and the result of the saturation will be to increase the population of level A and thus produce a greater population in level A than in level B. At the same time as this saturation is taking place a radiofrequency is now applied to the sample with a frequency such that $h\nu_{rf}$ is equal to the splitting between A and B. This then stimulates transitions from A to B, and the populations of the two levels will return to their normal equilibrium values. As a result the saturation of the electron resonance transition will be removed, and a strong ESR line will be suddenly obtained in place of the weakened saturated condition. The net result is that, if the detecting system is kept set on the electron resonance signal, a sudden increase in this will be obtained when the nuclear resonance signal sweeps through the condition:

$$h\nu_{rf} = \tfrac{1}{2}A - g_N\beta_N B \qquad (4.8)$$

It can also be seen that a similar situation arises when the radio frequency sweeps through the resonance value corresponding to the nuclear transition between levels C and D. Saturation of the electron resonance will have reduced the number of atoms in level D, but when the nuclear resonance transition is induced by the radio frequency field the population of levels C and D will be more or less equalised, and as a result the electron resonance signal being observed will again suddenly become desaturated and a large signal will be obtained. As a consequence, if the radio frequency signal is swept through a range of values, centred on $\nu = A/2$, a large increase in the electron resonance signal will be obtained when the frequency satisfies either of the conditions given by:

$$h\nu_{rf} = \tfrac{1}{2}A \pm g_N\beta_N B \qquad (4.9)$$

It is obvious that these two resonance conditions will give accurate values for both A and g very directly. It is also clear that the precise experimental requirements for ENDOR require that the applied d.c. magnetic field and the microwave frequency are held constant throughout the experiment at the resonance condition, but, at the same time, a varying nuclear resonance frequency is applied, and the values of its frequency, at which desaturation of the electron resonance signal is obtained, are carefully noted.

The power of this method can probably be best illustrated by one of the early examples in which ENDOR was applied to the study of F-centres in an irradiated KCl. The spectrum obtained is shown in *Figure 4.9*, and is typical of a double resonance tracing of this type.

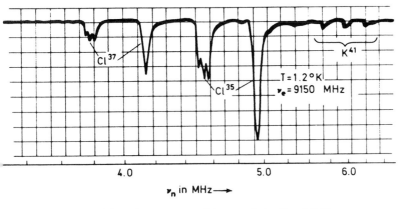

Figure 4.9. Endor spectrum of F-centres in irradiated K Cl

The ordinate represents the actual absorption produced by the electron resonance transition, with both the microwave frequency and the applied magnetic field held constant, as explained above. The abscissa corresponds to the changing value of the applied radio-frequency signal, and the absorption lines are obtained whenever this frequency corresponds to the actual hyperfine splitting present in the overall energy level pattern. It will be noted in this connection that the position of the lines correspond to *splittings*, and hence no symmetry is to be expected in the pattern as it is traced out. Since the factors which determine whether or not the desaturation is to occur are those which determine the width of the nuclear resonance line, it can be seen that the effective resolution has been increased by a factor of 1000, and this spectrum is an excellent example of the very great increase in resolution that can be obtained by the ENDOR method, since none of the hyperfine structure from either chlorine, or ^{41}K, could be observed in the straightforward ESR spectrum.

Although only the Overhauser and ENDOR techniques have been discussed in detail here, it should be mentioned that other types of double resonance can also be carried out, and a very good example of these is the double resonance which employs optical and radio-frequency radiations, and has developed into a large field in its own right on the radio-frequency spectroscopy of excited atoms. In all such types of double resonance one of the incident radiations is used to disturb the energy level populations corresponding to transitions which produce the other type of radiation, and in this way a sensitive detector is provided for the effect of one type of radiation upon the other.

Chapter Five

Applications of Spectroscopy in the Radio and Microwave Regions

5.1 Pure Research and Practical Applications

In this chapter a summary is given of the different ways in which spectroscopy at radio and microwave frequencies can be applied to obtain information of various kinds, and correlated with other branches of research in physics, chemistry and biology. Some of the applications have already been mentioned in earlier chapters especially those associated with the evaluation of fundamental physical and chemical data, such as the determination of nuclear spins and chemical bond lengths or angles. Other applications have been mentioned in passing, such as the possibility of extremely precise frequency standards in the microwave region, the accurate determination of magnetic field strengths, and the work that led on to the maser and laser. In a very general way these different applications can be grouped under two main headings, i.e., (i) determination of fundamental data and (ii) practical applications. *Tables 5.1* and *5.2* attempt to summarise the different applications of spectroscopy in these regions under these two main headings but it is, of course, obvious that such a grouping can only be very approximate and a good case can often be made for transferring the given application from one grouping to another.

In *Table 5.1*, where the determination of fundamental data is summarised under a series of different headings, the general principle has been to move from the smallest unit upwards, and thus the determination of nuclear properties are those first listed. Although most of these determinations of nuclear parameters have now been carried out, it is interesting to note that spectroscopy in these regions of the spectrum has contributed a very great deal in the way of fundamental data

Table 5.1

THE DETERMINATION OF FUNDAMENTAL DATA

	By gaseous spectroscopy	By electron resonance	By nuclear resonance
1. Nuclear Spins	Indirectly from h.f.s. components	Directly from number of h.f.s. components	Indirectly from intensities
2. Nuclear magnetic moments	Very indirectly via quadrupole coupling	Fairly directly from h.f.s. splitting	Directly from the resonance condition
3. Nuclear quadrupole moments	Directly from h.f.s. splitting	Indirectly from h.f.s. shifts and second-order transitions	Directly from nuclear quadrupole resonance experiments
4. Bond lengths and angles	Moment of inertia, I, determined directly, hence bond lengths, angles from isotopic masses	Resonance frequencies give g-value directly, hence electronic configuration and orbitals	Electric field gradients from quadrupole resonance. Diamagnetic nature of bond from chemical shift
5. Chemical analysis	Direct from spectra, but limited applicability	From h.f.s. patterns and sometimes from 'g'-values	From 'chemical shifts' and 'spin–spin coupling' parameters
6. Structural analysis	Via determinations of moments of inertia	From asymmetry of g-values, and 'spin–labelling' technique	From 'spin–spin' coupling parameters, and variation of spectra with temperature
7. Kinetic studies	In flash-photolysis and related studies	By 'continuous flow' and 'sudden freezing' techniques	Limited application because of lower sensitivity

Table 5.2

PRACTICAL APPLICATIONS

	By gaseous spectroscopy	By electron resonance	By nuclear resonance
1. Measurement of Frequency	Very accurate standards directly from absorption lines	Not applicable directly	From hyperfine splittings on atomic beams
2. Measurement of Magnetic Fields	Not applicable	In the 2 - 100 gauss $(2.10^{-4} - 10^{-2}$ tesla) region	In the 100 - 25 000 gauss $(10^{-2} - 2.5$ tesla) and below 2 gauss $(2.10^{-4}$ tesla) regions
3. Detection of Impurities	Very sensitive but rather restricted range of compounds	Sensitive and highly effective if unpaired electrons associated with impurity	Not very sensitive
4. Study of Irradiation Damage	Not applicable	By resonance from trapped electrons, interstitial atoms, or broken bonds	From change in line widths and slopes
5. In Radio-astronomy	Several molecules now detected in outer space	Not applicable except *via* study of Zeeman effect to give magnitude of inter-stellar magnetic fields	

on nuclear properties, and the very high resolution available has been a very important feature in obtaining such accurate information.

The next three groups listed in this table correspond to information of an atomic, or molecular kind and include the specific data on chemical bond lengths and angles, which can be deduced immediately from gaseous spectroscopy; together with the other two groupings of chemical analysis of a more general kind, both in the form of identification of different chemical species, on the one hand, and the more precise structural analysis, giving the relative position of individual atoms, on the other. The last group listed in this table covers the different kinds of kinetic processes that can be investigated by spectroscopy in these regions. There are, of course, many other items of fundamental data that can also be deduced from spectra in these regions, but this grouping under seven different headings may serve to summarise the general types of data that can be deduced, and they are each considered in somewhat more detail in the following pages of this chapter.

In contrast, *Table 5.2* groups the practical applications of spectroscopy in these regions of the spectrum under five headings, some of which are fairly specific and others somewhat broader. The first two concern the precise measurement of frequency, and of magnetic field, and follow immediately from the various techniques that have been described in earlier chapters. The third group, concerned with the detection of impurities, is really another way of discussing the high sensitivity in qualitative and quantitative analysis that is available in these forms of spectroscopy. Specific examples of these are of some interest, however, and do illustrate the particular advantages of spectroscopy in these different regions. In a similar way studies of radiation damage form a particular application of ESR spectroscopy, and are given a separate heading. Finally the various applications in radio astronomy have been listed, although these might just as well have been included in *Table 5.1*. The identification of various molecular groupings and fragments in outer space is proving one of the more fascinating studies of spectra in this region, and it does serve to link both the practical and fundamental sides of these studies in a fascinating way.

Each of these different applications is now considered in more detail, following discussion of the methods of determining fundamental data, mentioned before.

5.2 Determination of Nuclear Spins

It has already been seen that the nuclear spin, I, can interact directly with the unpaired electron of a paramagnetic substance to produce $(2I + 1)$ equally intense hyperfine lines in an electron resonance spectrum. It follows that electron spin resonance is probably the most direct and unambiguous method for the determination of nuclear spins, and, if the interaction is only with one nucleus, then the nuclear spin, I, is given immediately by equating the number of observed hyperfine lines to $(2I + 1)$. When the unpaired electron is interacting with more than one nucleus the hyperfine structure can become more complicated, as has been discussed in some detail in Sections 3.1 and 3.4. However, even in these cases, it is usually relatively simple to identify which features of the hyperfine pattern are to be associated with which particular nucleus, and thus determine the nuclear spin directly from the observed number of hyperfine lines.

It should be realised, however, that this technique can only be applied to nuclei which are present in a paramagnetic substance, and thus it may not have as wide an applicability as nuclear magnetic resonance, on the one hand, or gaseous spectroscopy on the other. It should also be noted that one of the basic requirements for the observation of such hyperfine splittings in electron resonance is high resolution, and thus narrow linewidths are required for the spectral lines. This often necessitates diluting the material with isomorphous diamagnetic substances, if the compound is to be studied in the solid state, or observing it in dilute solutions if studies are being made on liquids. In the case of some of the transition group atoms it is also sometimes necessary to cool the specimen to low temperatures to obtain narrow lines, because of the shorter spin–lattice relaxation times at the higher temperatures of observation. If, however, the spectra can be observed, then the determination of the nuclear spin, I, is normally quite unequivocal.

Although it might appear at first sight that nuclear magnetic resonance would give a more direct method of measuring the nuclear spin than electron spin resonance, the reverse is actually the case. This is due to the fact that the nuclear resonance absorption arises from transitions between successive levels formed by orientation of the nucleus in the magnetic field and, if no second-order effects are present, the absorption lines produced by successive transitions will, in fact, overlap, and be equivalent. This is illustrated in *Figure 5.1*, where the case of a nucleus with spin $I = 3/2$ is illustrated. Although there are three separate transitions arising from the four possible orientations of the spin of $3/2$ in a magnetic field, there will, in the simple case, only be a single absorption line since these three transitions all overlap completely. However, if second-order effects are present a slight shift in these energy levels may result, and in this case separate transitions will be produced and the nuclear spin will be able to be deduced from the

135

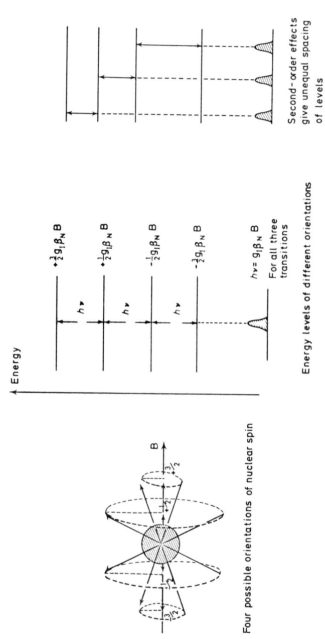

Four possible orientations of nuclear spin

Energy levels of different orientations

(a)

Second-order effects give unequal spacing of levels

(b)

Figure 5.1. N.M.R. for nucleus with I = 3/2: (a) If no second-order effects present the successive transitions exactly overlap; (b) If second-order effects are present separate transitions are produced giving 21 component lines

resultant pattern, the number of the resulting lines being, in general, equal to $2I$. If no second-order splittings are present then it is possible, in principle, to arrive at an estimate for the nuclear spin from the absolute intensity of the absorption line that is observed. This is, however, nothing like as unambiguous as the other methods described above, and the actual value of the spin, I, is not a parameter than can necessarily be deduced immediately from a nuclear resonance spectrum.

The hyperfine structure which is observed in the microwave spectroscopy of gases arises, not from an interaction between the magnetic moment of the nucleus and the surrounding magnetic field, but rather from an interaction of the electric quadrupole moment of the nucleus and the surrounding inhomogeneous electric field. This interaction is not so direct, or simple, as the magnetic interaction of the previous two cases and has been discussed in some detail in Section 2.5. It was seen there that the equations governing the number and spacing of the hyperfine structure lines are nothing like as simple as the $(2I + 1)$ rule of electron resonance spectroscopy, and will depend on the J quantum number of the molecular motion, as well as on the value of the nuclear spin, I, itself. The analysis of the spectra and the derivation of value of, I, may therefore be quite complicated but, nevertheless, because of the very great precision available, there is usually only one set of values which can be made to fit the observed spectrum and hence I can normally be quite unambiguously determined.

It should be noted that this method can only be applied to a nucleus in a gas molecule which has rotational transitions in the microwave region and, moreover, it cannot be used for nuclei with spins of $I = \frac{1}{2}$ or 0, since such nuclei do not possess quadrupole moments. As in the case of electron resonance, the crucial experimental condition is the necessary resolution in the spectra and this normally necessitates working at low gas pressures and avoiding saturation broadening.

5.3 Determination of Nuclear Magnetic Moments

The most direct determination of the magnitude of the magnetic moment of a nucleus comes from the resonance condition of the nuclear resonance absorption line. If the nucleus has a spin of $I = \frac{1}{2}$ the separation in energy between its two spin orientations will be equal to its magnetic moment multiplied by the strength of the applied magnetic field, and since the latter is known the former can be evaluated by equating this energy difference to the quantum of radio frequency which produces resonance absorption. If the nucleus has a spin greater than $I = \frac{1}{2}$, then the nuclear spin and the magnetic moment will be able to take up intermediate orientations in the direction of the applied field, as indicated in *Figure 5.1*, and the value of the total magnetic moment will then be obtained by multiplying the value, as deduced above, by $2I$, which is the number of equivalent transitions included within the single observed absorption line.

Relative magnitudes for nuclear magnetic moments can be determined directly from the hyperfine structure on an electron resonance spectrum. In this case the actual separation in magnetic field, between successive components of the hyperfine pattern, is directly proportional to the nuclear magnetic moment which is producing the hyperfine splitting. If hyperfine splittings, arising from an interaction with other nuclei, but coupled by the same electron wavefunction, are present in the spectrum, it is then possible to make an absolute measurement of the unknown nuclear magnetic moment by multiplying the relative hyperfine splittings by the value of the known magnetic moment. On the other hand, if no such direct comparison is available within the spectrum, the absolute value of the magnetic moment can only be deduced if the magnitude of the coupling constant between the nucleus and the unpaired electron is also known. This often involves somewhat complex molecular orbital calculations and hence the quantitative magnitudes will be approximate to that degree. It is for this reason that most determinations of nuclear magnetic moments, via electron resonance hyperfine splittings, have been carried out by the relative comparison method, and in this case very accurate values can be obtained.

As mentioned before, the hyperfine interaction which produces splittings in gaseous spectroscopy, arises from an electric rather than magnetic effect, and hence determination of nuclear magnetic moments is not normally made via microwave gaseous spectra. However, it is possible to evaluate such a nuclear magnetic moment from gaseous spectra if the Zeeman splitting of the hyperfine structure can be observed.

This interaction is only a second—order effect, however, since it acts via the main quadrupole interaction, and it also depends on the electronic wavefunction associated with the molecular binding. Again, therefore, direct measurements of nuclear magnetic moments are normally only possible by a comparison method and not in any absolute way.

5.4 Determination of Nuclear Quadrupole Moments

In the same way that accurate values of nuclear magnetic moments can only be obtained from magnetic resonance experiments, accurate values of nuclear quadrupole moments, which measure the distortion of the nucleus from spherical symmetry, can only be obtained from experiments in which the interaction is via the electric field; in other words, either from microwave gaseous spectroscopy or pure quadrupole nuclear resonance. The interaction of the quadrupole moment with the gradient of the molecular electric field has been discussed in some detail for hyperfine structure in gaseous spectroscopy in Section 2.5. It is clear, from a reference to equation 2.11, that a value for the product of the nuclear quadrupole moment and the gradient of the electric field can be derived directly from the frequency of the observed spectral lines. It is these 'quadrupole coupling constants' which are therefore determined quantitatively, but relative quadrupole moments themselves can be determined very accurately. Absolute values of the actual moment, Q, however, do depend on a knowledge, or calculation, of the electric field gradient at the nucleus in question.

The same kind of general comment also applies to measurements obtained from pure quadrupole nuclear resonance. In this case no external magnetic field is applied to the specimen, but instead the frequency of the incoming radiation is adjusted, so that its energy is equal to the splitting produced between the levels which correspond to different orientations of the nucleus in the internal electric field gradient. The nuclear quadrupole moment, or coupling constant, can then be determined directly from the resonance frequency at which the absorption occurs.

5.5 Chemical Bond Lengths and Angles

The determination of bond lengths and angles is one of the basic objectives in the study of structural chemistry, and this information can be deduced in a variety of ways from spectroscopy in the microwave and radio regions. Probably the most direct method is via the absorption frequencies or rotational lines in gaseous microwave spectroscopy. It has already been seen, from the various equations quoted in Sections 2.1 and 2.3, that the frequency of the spectral lines can be related directly to the moments of inertia of the molecules concerned, and if the masses of the nuclear isotopes are accurately known, as is usually the case, these moments of inertia will in turn give the distance between the nuclei, and hence the chemical bond length, very accurately. In a similar way the angles can also be determined by measurements on moments of inertia around more than one axis, and the basic structure of the molecule can thus be resolved.

Values for internuclear distances, and thus of chemical bond lengths, can also be deduced from solid state nuclear resonance measurements. It has been noted earlier that the line width and shape, and sometimes the line splitting, of spectra obtained from single crystals can be related directly to the spin-spin dipole-dipole variation which has the general form $(3\cos^2\theta - 1)/r^2$. The actual magnitude of the broadening, or splitting, obtained can thus be related directly to the inverse square of the nuclear distances. The information obtained in this way is nothing like as precise as that obtained from the absorption line frequencies of the gaseous spectra, however, and, since the vast majority of work now carried out on nuclear resonance is on high resolution spectra in solution, this is no longer very extensively used as a method of obtaining such data.

Although bond lengths, as such, do not normally come directly from measurements on electron spin resonance spectra, it is nevertheless often possible to deduce a great deal of information about the chemical binding from the g–value of the observed resonance, and its angular variation when studied in a crystal. These g–value magnitudes effectively measure the amount of orbital angular momentum admixed with the spin momentum, and this, in turn, is determined in a very precise way by the chemical bonding in which the atom under study is taking place. Hence the precise nature of the chemical bonding can often be deduced from the g–value measurements, which often give structural information about the surroundings of the paramagnetic atom itself. Specific examples of this are taken up in Section 5.7, which discusses structural analysis in more detail.

5.6 Qualitative Chemical Analysis

As in all forms of spectroscopy, qualitative analysis can normally be effected by a direct measure of the observed absorption line frequencies, and a correlation of them with the known values for given compounds. This certainly applies to all normal microwave gaseous spectra and also, as has already been considered at some length in Chapter 4, to the higher resolution spectra of nuclear magnetic resonance. In this last case it is not only the particular atom which can be detected, but the chemical groups in which the atom is situated and, as the tables in Chapter 4 have already indicated, very precise differentiation between different chemical groups can be obtained in this way. The direct identification of different chemical groupings arises from the different values of the chemical shift observed in such spectra, but further information on the actual spacing of groups, relative to each other, is obtained from the spin–spin coupling measurements which can be made on the same spectrum.

Unambiguous characterisation in this way is not really quite so readily available from electron resonance spectra, because any given paramagnetic atom, or complex, may well cover a range of g-values, and precise identification from a g-value measurement alone is often not possible. However, the observation of hyperfine structure on the spectra is frequently of very great assistance in this connection, and the characteristic $(2I + 1)$ equally intense hyperfine lines that are produced from nuclei with magnetic moments can often be used as a very powerful method of qualitative and quantitative analysis.

One relatively new application of electron resonance should be mentioned in this connection, however, and this is the 'spin-labelling' technique. This can be used to analyse the composition of large individual molecules, and also to act as a kind of alternative radio tracer technique, so that labelled molecules can be followed through a human body or other reacting system.

The basic idea is to attach a well-characterised free-radical group to a large biochemical or biological molecule, so that the whole molecule then becomes 'labelled' with an unpaired electron. The presence of these molecules can then be detected later, in very small concentrations, by a sensitive e.s.r. spectrometer and all the ideas developed for radioactive assay can be employed. It has wider application and none of the radiation limitations of radioactive tracers, however, and has already been applied to human investigations, such as the tracing of various drug concentrations through the metabolic system.

In addition to acting as a tracer technique, the spin-labels can be used to probe their immediate surroundings on the molecule itself. Thus most of the spin-labels used have an NO group included with them, with the nitrogen atom bound to a tertiary carbon atom. The structure of such a free radical is shown in *Figure 5.2*, with the unpaired spin distribution indicated over the nitroxyl group bonds. It is seen that the major part of this will be located in the p-orbit of the nitrogen atom, and this asymmetry in space is reflected by the anisotropy of the observed g-values which vary from 2.002 to 2.009 parallel and perpendicular to the p-orbital axis. The hyperfine splitting between the three lines from the nitrogen interaction are similarly anisotropic, and these well-characterised features of the spin-label allow its orientation to be deduced from the observed e.s.r. spectra.

It is thus possible to affix the spin-label at different points of the biochemical molecule, for example, along successive steps in a polypeptide chain, and determine the orientations of it, and, hence of the associated part of the molecule, at these various positions. In some

141

cases the binding to the spin-label is sufficiently weak to allow free rotational motion. If this is so, and its immediate surroundings do not

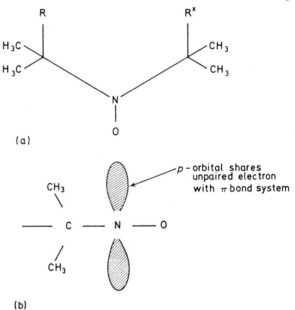

Figure 5.2. Structure of typical 'spin-label': (a) Basic structural formula. R and R^x can be a variety of different groups and are used to bind the spin-label to the protein; (b) Unpaired electron density residing mainly in π - orbital association with p-orbit on nitrogen atom

preclude such rotation, a motional narrowing of the lines will occur. The existence of these narrow line 'mobile' spin-labels can then be used to give precise information on the openness of that particular part of the molecule.

It is clear that these 'spin-labels' can be used in a variety of ways as a tracer technique, or for qualitative and structural analysis, and they have opened up a whole new range of applications of electron resonance spectroscopy.

5.7 Structural Analysis

5.7.1 From N.M.R. and Gaseous Spectroscopy
It was pointed out in the last section that the combination of measurements of chemical shift, and spin–spin coupling, on nuclear resonance

spectra can give detailed information on the relation of groups within the molecule, as well as a simple analysis of the constituent groups themselves. Information on structural analysis is also available from the determinations of moments of inertia via normal microwave gaseous spectra, but in the molecules normally studied the structure is often known to be either linear, or of a symmetric or antisymmetric top type, and no major unknown questions of structure need to be solved in the first place.

5.7.2 From g-value Variations in E.S.R.

The application of electron spin resonance to structural analysis takes place in a somewhat more indirect manner, but can nevertheless often give vital information which is not accessible by other means. As one example of this, the case of the electron resonance spectra of the haemoglobin molecule might be quoted. The structural form of the haemoglobin molecule is represented schematically in *Figure 5.3*, from which it is seen that a single iron atom is located at the centre of a plane, and is surrounded by four nitrogen atoms. The plane is called the 'haem' or porphyrin plane, and the rest of the protein in the molecule is attached as a series of coiled polypeptide chains below the plane, via the nitrogen atom of a histidine ring. The sixth coordination point of the iron atom is occupied by an oxygen molecule when the haemoglobin has just passed through the lungs and become oxygenated, or alternatively, by a carbon dioxide molecule after the oxygen has been used by the body. Other atoms, or groups, can also be attached to this sixth coordination point, and when the haemoglobin is crystallised from suitable buffer solutions a water molecule is normally present.

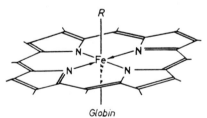

Figure 5.3. Structural form of centre of hae-moglobin molecule. The iron atom resides at the centre of the haem or porphyrin plane at the centre of a square of four nitrogen atoms

As will probably be appreciated, there are some groups, such as CN, which, once attached to the iron atom, are impossible to remove. The electron resonance investigations on haemoglobin were initiated to find out as much detail as possible about the bonding between the iron atom and the different groups on the sixth coordination point. This bonding must involve very precise balancing of the energies concerned, because the actual number of unpaired spins changes significantly when the oxygen is associated with the iron atom. Since this bond is formed and dissociated, by the simple process of breathing in and out, the energies cannot be of a very large magnitude. During the early investigations of electron resonance it became apparent, however, that electron resonance could also give very important information on the actual structure of the haemoglobin molecule itself.

The iron atom at the centre of the haem plane can be of either ferrous or ferric valency and the binding can also be classified approximately as ionic, or covalent, depending on whether the $3d$ orbitals are used by the electrons of the ligand atoms, or are left empty for the $3d$ electrons of the iron atom itself. In the case of the ionically bound ferric derivatives there will be five unpaired electrons left in the $3d$ shell, and there will be five vacant orbitals for them to fill. They will therefore all align, with their spins parallel, to produce a resultant $S = 5/2$, which is a case identical to the well-studied Mn^{2+} inorganic compounds. Three degenerate doublets are therefore expected in a solid state crystal, these corresponding to possible orientations of the $S = 5/2$ resolved along the axis of the crystalline electric field with $M_s = \pm 1/2$, $\pm 3/2$ or $\pm 5/2$. In simple inorganic compounds the splitting between these three doublets is small, compared with the size of the microwave quantum, and in these cases transitions between all three doublets are observed, and the resultant electronic splittings give rise to five sets of electron resonance lines for each ferric, or manganese, atom. In the case of the haemoglobin derivatives, however, the energy splitting between these three doublets is very much greater than the energy of the microwave quantum used, and, as a result, only a single electron resonance line is observed corresponding to the $S = \pm 1/2$ doublet, which lies as the ground state.

This ground level no longer has an isotropic g-value equal to 2.0, however, but a g-value which varies from 2.0 along the axis of the internal molecular electric field, which in this case is the normal to the haem plane, to a value of 6.0 in the direction perpendicular to this axis, i.e., for any direction lying within 'haem' plane itself. The actual spatial variation of this g-value variation for the haemoglobin molecule is shown in *Figure 5.4,* and it is evident that it provides a method for accurately locating the orientation of the haem planes. When these electron resonance investigations were carried out, the X-ray studies had not been completed and it was possible to determine the orientation of the haem planes quite precisely by the electron resonance measurements, and then hand this information to the X-ray crystallographers, to assist in their complete analysis of the rest of the molecule.

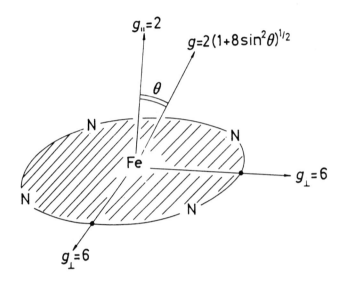

Figure 5.4. Spatial variation of g-value for haemoglobin

In principle all that is required to locate the orientation of the haem planes, is to mount a crystal in the cavity resonator and rotate it in all possible directions until a g-value equal to 2.0 is obtained. The magnetic field is then being applied along the direction normal to the haem plane, and hence its orientation is determined. In practice it is very difficult to move a crystal in three directions at once, inside a cavity resonator, and the crystals are therefore mounted in different crystallographic planes, in turn, and the g-value variation for the different planes is plotted out.

Such g-value variations for three crystallographic planes of a myoglobin crystal are shown in *Figure 5.5* and, in this particular derivative, the crystal is of monoclinic form with two molecules per unit cell. Myoglobin transports oxygen in muscle fibre, as opposed to the blood stream, and each myoglobin molecule only contains one iron atom, and associated haem plane, and hence only two resonance lines are to be expected, one corresponding to each molecule of the unit cell. The g-value variations for the two different molecules can be clearly seen in *Figure 5.5*, and the crossover points in this figure locate the directions

145

of crystallographic *a* and *b* axes quite precisely. It is clear from a consideration of the g-value variation shown in *Figure 5.5* that the quantitive results of the *g*-value variation in these three planes will enable the precise orientation of the haem planes to be determined, as is shown in *Figure 5.5(d)*.

If similar measurements are carried out on crystals of haemoglobin, where each molecule contains four iron atoms and haem planes, the

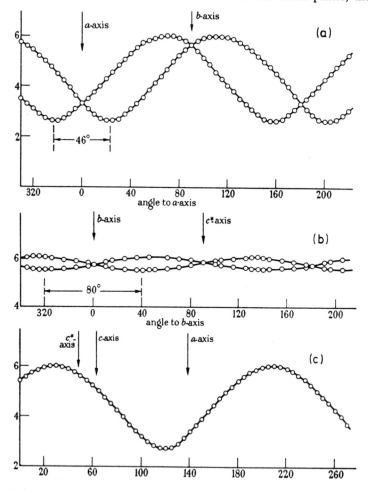

Figure 5.5. g-value variations and haem plane orientations deduced from them; (a) g-value variations in ab crystallographic plane of myoglobin; (b) Variation in bc plane; (c) Variation in ac plane

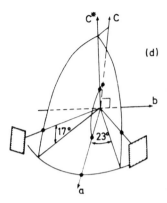

Figure 5.5. (d) Orientation of haem planes deduced from g-value variations

results obtained will be as shown in *Figure 5.6*, and these can again be analysed in the same way as before, to give the precise orientations of the four 'haem' planes per unit cell, as shown in *Figure 5.6(b)*. These determinations of the precise orientations of the haem planes within the myoglobin and haemoglobin molecules serve as very good examples of the potentialities of electron resonance in the investigation of compounds of biochemical interest, and illustrate the way in which structural information can be so deduced. It will be appreciated that in these particular studies the g-value variations have been used simply as a probe of the symmetry surrounding the iron atom. It has not been necessary to invoke any detailed theory on molecular orbitals themselves, and the whole argument has been based on one of pure geometrical symmetry.

Additional information can often be obtained if further measurements are made on different derivatives, and if molecular orbital theory is applied to analyse second-order effects. A good example of this is the measurements made on the myoglobin and haemoglobin covalent derivatives, such as the complex in which three nitrogen atoms are linked to the sixth coordination point, in place of the water molecule or oxygen atom. These then produce derivatives which have a resultant spin of $S = \frac{1}{2}$, and one has to consider the motion of this single unpaired electron in the d-orbitals of the iron atom, as they are affected by the surrounding p and σ-orbitals of the nitrogen atoms. Therefore the detailed arguments and reasoning that has been developed in Section 3.4 and illustrated in *Figures 3.8, 3.9* and *3.10*, and which can

147

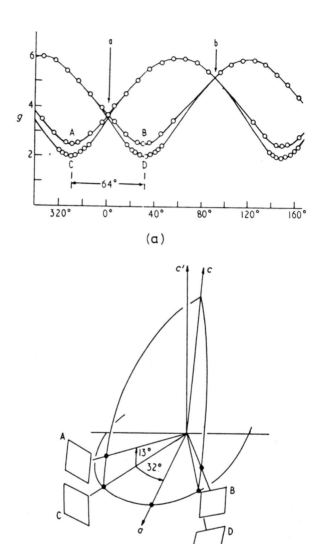

Figure 5.6. g-value variations and haem plane
orientations for haemoglobin: (a) g-Value
variation showing the four haem planes per
molecule; (b) Haem plane orientations

be redrawn for the particular case of the haemoglobin, as in *Figure 5.7*, can be applied.

Thus, a theoretical analysis of the observed g-values, as illustrated in the figure, show that they are only consistent with an energy level

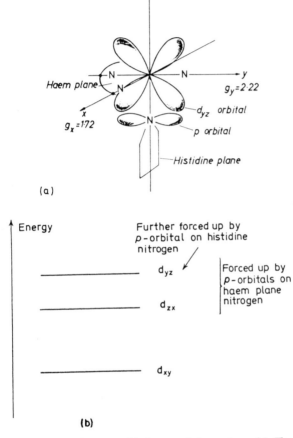

(a)

(b)

Figure 5.7. Electron orbitals around iron atom: (a) The interaction of the p-orbital of the nitrogen of the histidine plane raises the energy of the closest d-orbital – in this case d_{yz}; (b) Resultant energy level splittings

diagram in which the d_{xy} orbital must lie lowest, with the d_{zx} and d_{yz} higher in the group. The particular order of these energy levels may be explained by considering the interaction of the magnetic d-electron, on the iron atom, with the p-electron density on the surrounding nitrogen atoms. As explained in Section 3.4 the fact that the energy level of the d_{yz} orbit is noticeably higher than that of the d_{zx} orbit indicates that an interaction between its magnetic electron and the p-orbit of the nitrogen on the histidine ring must indeed be taking place. Thus, the direction of the histidine plane is determined as parallel to the x-axis of the g-value variation, as indicated in the figure.

The main step in this additional analysis has been to use the asymmetry of the g-value variation in the haem plane as a probe of the asymmetry in the structure of the molecule below the plane, and this illustrates rather nicely the way in which second-order effects can also be used to determine structural information in such studies as these.

5.8 Kinetic Studies

Several examples of the study of rapidly changing transient systems have already been quoted in earlier chapters of the book. In particular it was noticed in Section 3.7 that electron spin resonance can now be employed to follow rapidly changing situations in relation to both enzyme systems and photochemical reactions. *Figure 3.15* is a very good example of the power of such transient studies not only to identify the different transition group atoms, or free radical intermediates, which are taking place in the interaction, but also to follow very accurately the kinetics of the different steps.

It should also be noted, however, that some of the earliest such measurements on ESR were used for the study of the triplet state, which can be regarded as a specific type of photochemical reaction. The principles underlying the energy levels of a triplet state have been discussed in some detail in Subsection 3.3.4, when considering electronic splitting in general terms. Reference back to *Figure 3.7(b)* will remind one that when a molecule containing two unpaired electrons, which are coupled to give a resultant $S = 1$ (as in the case of a triplet state), exists inside a crystal then the internal electric fields will split the energy level system to produce two separate transitions, as indicated in that figure. Moreover, the interaction with the field is likely to have the normal $(3\cos^2 \theta - 1)$ angular variation and it follows that the electronic splitting between the $M_S = 0$ and the $M_S = \pm 1$ levels will also vary with the angle, as indicated in *Figure 5.8*. Thus, there are not only two separate resonance absorption lines produced for each angle, but the separation between these varies with the angle, as shown in the diagram. Hence, if glassy, or polycrystalline, samples are studied, with random orientation of their molecules, a complete range in the values of Δ is obtained, and

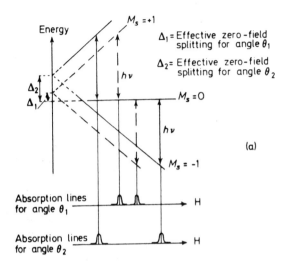

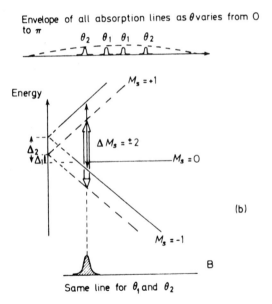

*Figure 5.8. Variation of electronic splitting with angle:
(a) For the normal $\Delta M_S = \pm 1$ transitions, a variation in
the magnetic field spacing of the two resonance lines is
produced which smears out the resultant absorption; (b)
The two $\Delta M_S = \pm 2$ transitions will always overlap, however, and hence a single isotopic line will be obtained*

151

the overall absorption is smeared right out along the field axis, as indicated in the figure. It is therefore not suprising that all the initial efforts to detect triplet states by electron resonance ended in failure since most of these were made on liquids or low temperature gasses. Thus, the observation of the main transition in a triplet state is only likely to be successful if careful measurements are made on a crystal lattice which has been diluted, if possible, to reduce line broadening effects.

The first successful detection of triplet states by electron resonance was achieved under exactly these kinds of conditions i.e., on single crystals of durene containing a small amount of naphthalene, which was irradiated with light from a high-pressure mercury arc light. The crystal was mounted on plastic wedges, cut at different angles, so that the angular variations of the spectra could be studied in different crystallographic planes. It was then found that a doublet spectra of the type predicted was indeed present and varied with angle very markedly.

A further major step forward in triplet state studies came, however, when it was realised that it is also possible to observe $\Delta M_s = \pm 2$ transitions in some cases. These would be forbidden on simple first-order theory, but second-order effects can produce quite a significant transition probability. The great advantage of studying these transitions is that they take place between the highest and lowest of the three electronic levels as indicated in *Figure 5.8(b)*. Their resonance field value is therefore not altered to any great extent by the angular variation of the zero field splitting, and they should consequently be detectable from liquid or solid solutions, or other media with a random orientation of the molecules. Since the energy transition is twice as great as normal it also follows that the field value for resonance is about half of that normally expected.

Such signals were, in fact, first obtained from naphthalene in a rigid glass solution in glycerol, after radiation by ultraviolet light at 77K. Following these initial measurements a large amount of work has now been carried out on the triplet state, and the ability to study solid solutions, rather than crystals, does give conditions which are much closer to those normally obtained in practice.

5.9 Practical Applications — Measurement of Frequency

The applications of spectroscopy at radio and microwave frequencies so far considered have been those that provide information of a fundamental nature concerning nuclear and atomic, or chemical, properties. As has been noticed through the earlier chapters, however, an increasingly large number of practical applications have followed from work in these regions of the spectrum and a very brief summary of some of these is now attempted, beginning with the application to frequency standards and the measurement of time.

Since the resonance frequencies of gaseous molecular spectra are determined entirely by the interatomic binding forces, and not by any external conditions, they offer the possibility of very accurate frequency standards that can be reproduced at any place desired on or off the earth's surface. This does not apply to microwave or radio-frequency, spectral lines that are obtained by magnetic resonance methods, as those resonance conditions require the determination of a magnetic field strength as well. On the other hand, the hyperfine splitting of atomic spectra, and of atomic beams in particular, are also independent of any externally applied magnetic fields and these therefore offer an alternative atomic standard of frequency and time. Details of the way in which microwave gaseous spectra were first used to provide new standards of frequency and time, in the form of the ammonia maser, have been given in Section 2.6 and all succeeding microwave gaseous standards have followed along the same general principles.

In order to construct an atomic beam frequency standard some such element as caesium is employed, which has a hyperfine splitting, independent of external fields, in the region of 9000 MHz. The beam of caesium atoms is then fired down an evacuated space, past one set of magnets producing an inhomogeneous field in one direction, then through a microwave cavity, and finally through a second inhomogeneous magnetic field. The frequency of the oscillations in the microwave cavity is adjusted to cause a spin transition in the caesium atoms as they pass through the microwave radiation, so that their magnetic moments are reorientated and the atoms become deflected away from their normal path when passing down the second inhomogeneous field. Thus, on resonance, the output of the detector will be a minimum and any shift of frequency in the multiplier-chain feeding the microwave cavities will cause an increase in detected current, which can be used to derive an error signal and reset the master quartz oscillator. In this way the frequency of the driving circuits can be linked directly to the frequency of the hyperfine transition of the caesium atom. The basic principles of operation of such a caesium clock are illustrated in *Figure 5.9*. It can be seen that these basic principles are the same as those already discussed for radio-frequency spectroscopy of atomic beams in Section 1.6.

Considerable work has been carried out in recent years on the development of such standards of frequency and time, which cannot be dealt with in detail here, but the accuracy achievable with these atomic and molecular standards is now far greater than that available from the

153

older astronomical standards, and they have taken over in all accurate work such as the calculation of satellite flight paths and similar studies.

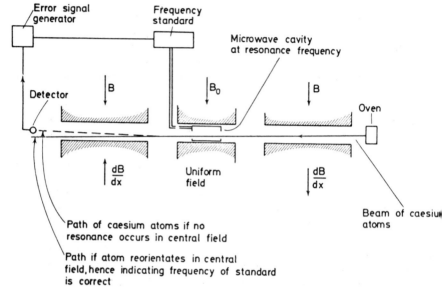

Figure 5.9. Basic principles of caesium atomic clock. The principles are the same as those previously illustrated in Figure 1.10. On resonance the caesium atoms are deflected away from the collector by the second inhomogeneous field. Hence any shift of frequency will cause an increase in detected current which will then produce a correcting signal

5.10 The Measurement of Magnetic Fields

Because the frequency of absorption of electron and nuclear resonance lines depends linearly on the strength of the applied magnetic field they offer a very direct method of magnetic field calibration, converting the determination of field strength into one of frequency. If a resonance absorption line is to be used for magnetic field measurements it should have as high an intensity as possible so that only small quantities need to be employed, and thus small regions of the field can be measured independently. The absorption line should also have a linewidth that is as small as possible, because this will be one of the limitations on the accuracy of measuring the frequency. Two compounds fulfilling these conditions reasonably are (i) for nuclear resonance, protons in water, with the spin–lattice relaxation time reduced somewhat by the addition of a paramagnetic salt and (ii) for electron resonance, crystals of the free radical diphenyl trinitrophenyl hydrazyl which has a very intense signal, with a strong exchange-narrowing; or, alternatively, as a secondary standard, samples of carbonaceous material, like anthracite which also produce a strong signal but with a somewhat broader line.

Whether one uses nuclear or electron resonance for the determination of a magnetic field depends on both the strength of the magnetic field to be measured and the frequencies available for the determination. It will be remembered that the strengths of the magnetic fields for proton and electron resonance are given respectively by:

$$B = 2.3487 \, \nu \text{ for protons}$$
$$B = 3.566 \times 10^{-3} \, \nu \text{ for free radical electrons}$$

where the field, B, is in tesla in both cases.

Accurately calibrated wavemeters are usually readily available in the region of 1 to 100 MHz and this corresponds to magnetic fields from 200 to 20 000 gauss (0.02 to 2 tesla) for nuclear resonance, and from 0.5 to 30 gauss for electron resonance; with the region in between 30 and 200 gauss covered by both methods. It should be noted that in both frequency regions it is possible to use much simpler equipment than that employed for full-scale resonance measurements, and, in particular, a proton resonance magnetic field meter can be made in the form of a simple marginal oscillator which will nevertheless produce a fairly intense signal from a specimen containing only ½ cm^3 water.

It has also been noticed in the previous chapter that nuclear resonance methods in the form of free-precessing induction signals can be employed to measure very small values of magnetic field, and in particular to determine the strength of the earth's magnetic field very accurately at any given point. Reference back to Section 4.7 gives a clear summary of this important new development, which is finding rapid application in such fields as geological and archaeological prospecting.

5.11 Detection of Impurities

In principle it should be possible to use any form of spectroscopy to detect impurities, since the impurity atoms or molecules will be identified by their characteristic spectral lines as distinct from the bulk of material in which they reside. The success of the particular form of spectroscopy employed will depend on two factors, the resolution, and the sensitivity, of the spectrometer system. The resolution must be such that the spectral lines of the impurity are not overlapped by the more

155

intense main lines present, and the sensitivity must be such that an observable signal, noticeably above the noise level, is obtained from the concentration of impurities that it is desired to detect.

So far as sensitivity is concerned, microwave gaseous spectroscopy has one great advantage over any form of magnetic resonance spectroscopy in that the coupling to the electromagnetic radiation is via the electric vector, which produces much higher transition probabilities than for coupling via the magnetic field. As a result, much smaller concentrations of impurity will produce observable signals in microwave gaseous spectroscopy than in magnetic resonance. This considerable advantage of gaseous spectroscopy is offset, however, by the somewhat restricted range of molecules that can be studied, and it is more than likely that any particular impurity, that it is necessary to detect, will not have rotational spectra in the microwave region. Apart from some very specific cases, therefore, microwave gaseous spectroscopy has not been used significantly for impurity detection.

When a comparison is made between the sensitivities of nuclear and electron magnetic resonance the crucial factor is the much larger energy spacing between the electron levels, for a given magnetic field, as reflected by the much higher resonance frequency that is required. This larger splitting produces a larger population difference between the ground and excited states, this difference being given by Maxwell-Boltzmann statistics as

$$N_1/N_2 = \text{esp} \ (-h\nu/kT) \tag{5.1}$$

and therefore

$$N_2 - N_1 \simeq N_2 \ h\nu/kT \tag{5.2}$$

As a consequence, at a given temperature, the population difference, and hence the strength of the observed signal, will be directly proportional to the frequency of the resonance radiation. This major difference between electron and nuclear resonance therefore suggests that the former will be some 2000 times more sensitive then the latter, and, in general, this is found to be the case. Thus, in broad terms, electron resonance spectrometers can detect down to 10^{11} atom per cm^3, while the normal sensitivity of nuclear resonance spectrometers is limited to about 10^{15} atom per cm^3.

Although the absolute sensitivity of nuclear magnetic resonance is limited in this way, it can still detect small relative proportions of impurities present in a sample, and the very high resolution obtainable in solution greatly assists in identification.

Most of the practical applications of impurity detection, however, are concerned with small absolute quantities, and usually in the solid state. In these circumstances electron resonance has considerable advantages and it has been used in quite a number of important cases. One of the first major applications in this field, was to the identification of small traces of doper atoms in semiconductor materials. Thus doper, or acceptor, atoms in semiconductor material will have an unpaired electron associated with them, having either released an electron to the conduction band or collected one from it. Hence they will give rise to an ESR signal, and this will have a hyperfine splitting imposed upon it, characteristic of the particular donor, or acceptor atom. As examples, phosphorus will produce a simple doublet from the $I = \frac{1}{2}$ of ^{31}P; arsenic a quartet from the $I = \frac{3}{2}$ of ^{75}As; while the two main isotopes of antimony give overlapping sets of six and eight lines. It is therefore simple to identify the particular impurity atoms present in this way, and the sensitivity is such that concentrations down to 1 part in 10^{12} can be detected.

5.12 Study of Irradiation Damage

The detailed study of irradiation damage is normally undertaken on solid specimens because the damage normally 'heals' rapidly in a liquid or gas. Although nuclear resonance has been used to study solids which have been subject to high energy radiation, these investigations do not have the high resolution of the liquid state. Hence most of the detailed studies on irradiation damage have employed electron resonance techniques, and these are particularly suitable for detecting and characterising such damage centres because they will normally have an unpaired electron associated with them.

5.12.1 Studies on Inorganic Crystals

Such studies can be grouped under two main headings i.e., (i) studies on simple inorganic crystals, and (ii) studies on organic compounds, and biological material in particular. The general principles behind the investigation of the inorganic compounds are very similar to those outlined in the last section for impurity atoms in semiconductors, the main identifying feature again being the associated hyperfine structure. This may be illustrated by a specific case, such as that observed from irradiated magnesium oxide. Irradiation will displace some of the oxygen atoms, and one then has a vacancy which will trap an electron, and this will be surrounded by six equally-distant Mg^{2+} atoms. Most of these have isotopes with no nuclear spin, and which will therefore produce no hyperfine structure. However about 10% of these will be ^{25}Mg, with a nuclear spin of $I = 5/2$, and which will therefore produce a six-line hyperfine pattern.

The actual ESR spectrum that is observed from irradiated MgO is seen in *Figure 5.10*, where the large central line comes from the vacancies which have no ^{25}Mg nuclei in the six atoms around the trapped electron. The smaller six-line hyperfine pattern, also present, comes from those vacancies which have one ^{25}Mg nucleus coupled to the unpaired electron; and the observed intensities confirm this interpretation precisely. This is a particularly simple and straightforward case, but it does illustrate the basic methods of analysis very clearly and these can be easily extended to the more complex damage centres that can also be studied in this way. It should also be pointed out that such studies as these are exactly the type that can benefit from the higher effective resolution available from ENDOR techniques. The particular example quoted to illustrate ENDOR in action in Section 4.8, and shown in *Figure 4.9*, is in fact just such a study of irradiated KCl.

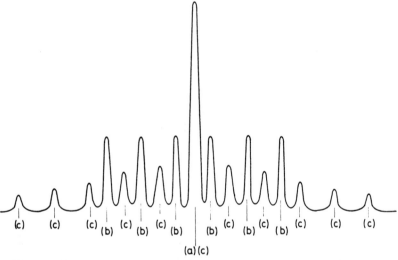

Figure 5.10. E.S.R. spectrum from irradiated MgO. The large central line (a) is due to centres with no Mg^{25} nuclei in the six atoms around the trapped electron. The six equally intense hyperfine lines (b) are due to centres with one Mg^{25} in the six atoms. The smaller eleven-line pattern with a binomial distribution in intensity (c) is due to centres with two Mg^{25} nuclei in the surrounding six atoms

5.12.2 Studies on Biological Materials

The other large group of irradiated compounds which have been studied extensively by ESR are those of a biological nature. Some of the very early ESR studies were undertaken on amino acids and proteins, and later work also showed how resultant changes in the material could be

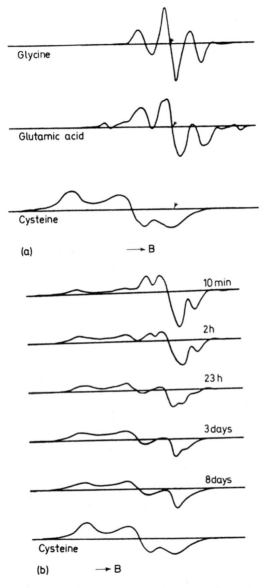

Figure 5.11. E.S.R. Spectra from irradiated proteins: (a) Spectra obtained from X-irradiated glycine, glutamic acid and cysteine; (b) Time variation of spectra from irradiated glutathione

159

followed, and linked with possible mechanisms for radiation protection. This is illustrated in *Figure 5.11*, where the spectra obtained from irradiated glycine, glutamic acid, and cysteine are shown on the left, while the time variation of a spectrum obtained from glutathione after irradiation, is shown on the right. It is clear that the cysteine spectrum is very different from the other two on the left, and this is because the S–S bond tends to localise the unpaired electron, which will then have significant spin–orbit coupling to the S atoms, and hence a shifted *g*-value.

If the variation of the glutathione signal is now studied, it is seen to change gradually after irradiation and become more and more like the cysteine signal. This suggests that the unpaired electrons formed during the irradiation are being gradually taken up by the S–S bond, which thus acts as a kind of 'sink' for these active electrons and inhibits the damage process.

A large amount of work on compounds of all kinds, treated by all types of radiation, has been undertaken, and the high sensitivity of the ESR technique, together with the characteristic hyperfine structure, has proved invaluable in such studies.

5.13 Applications in Radioastronomy

One of the most impressive and fascinating applications of spectroscopy in both the microwave and radio regions has come from outer space in recent years. For some time after its inception radioastronomy was mainly concerned with the 'white noise' or, whole range of frequencies, emitted by radiostars, or galaxies, and it was not until some years later that the first spectral lines were identified.

The first of these to be studied was the famous 1420 MHz, or 21 cm wavelength, line of atomic hydrogen. This particular frequency, and spectral line, arise from the fact that the isolated hydrogen atom can exist in two basic forms, i.e., with its electron and proton spins parallel, or antiparallel. There will be a small difference in energy between these two states, and the hydrogen atoms can be converted from one to the other by the emission, or absorption, of the appropriate resonance frequency, which is the 1420 MHz observed.

It was quickly realised that such measurements could be utilised to gain a large amount of information about our galaxy in particular, and the universe at large. Thus the plotting out of such atomic hydrogen concentrations enables the positions of the galactic arms to be delineated, and the relative motion of these can then be deduced from the Doppler shifts observed on the 1420 MHz frequency.

This first radio-frequency spectral line from outer space was, as its name implies, *atomic* in nature, but in 1968 molecular astronomy came into being with the discovery of OH radicals. One of the most striking features of these observations was that the hydroxyl spectral lines were often observed as intense narrow *emission* lines, rather than absorption lines, and indicated that 'maser' action, with microwave pumping, was taking place in intergalactic space. Further work then showed that the microwave spectra of several molecules, including ammonia, formaldehyde and carbon monoxide could be picked up from interstellar gas clouds near the galactic centre.

The number of such molecules being identified is increasing steadily year by year. These studies not only give information on the actual constitution of the galactic gas clouds, but also very significantly alters the overall picture we have of the nature of our own galaxy; both in the way of unexpected concentration of molecular species, and on the nature of the physical conditions that must be present to allow such phenomena as 'maser' action to take place.

This brief summary of the way in which spectroscopy at radio and microwave frequencies is now being used to probe the very large features of our Universe, such as the galactic arms, and their rotation and physical properties, seems a very appropriate point at which to bring this book to a close. It has been seen how these new types of spectroscopy have been applied to study a whole range of particles and interactions which stretch from nuclear properties at one end, through simple molecular features, to the specific properties of crystals and solids as a whole, and now, finally, of the galaxy itself.

Another general feature seen weaving continuously through all this work has been the constant interplay between the research of the pure scientists in the Universities or Research Institutes, on the one hand, and the work of the applied engineers, in Industry or Government Establishments on the other. Each has been helping the other to produce better tools and techniques, which have then been used to initiate new ideas which, in turn, led to further applications, and then to even better tools and techniques. So the wheel has continued to move full circle. It is on this positive note, and the belief that such work illustrates precisely the way in which science and technology should progress together, that a text on spectroscopy at radio and microwave frequencies should close.

Index

165